Zen of Stars

Futures of Planet Earth

St.Clair

2021

Argos

Window Photograph © 2005 by St.Clair
Chillon Castle Photograph © 2005 by St.Clair
Back Cover Photograph of St.Clair © 2002 by Paul Davis

Cover photo – The Sombrero Galaxy, M104 Constellation Virgo
Credit: NASA/JPL-Caltech and The Hubble Heritage Team
(STScI/AURA)
Argos Publisher in association with Trafford

Note for Librarians: A cataloguing record for this book is available from Library and Archives Canada at www.collectionscanada.ca/amicus/index-e.html
ISBN 1-4120-9088-1

PUBLISHING™

Offices in Canada, USA, Ireland and UK

Book sales for North America and international:
Trafford Publishing, 6E–2333 Government St.,
Victoria, BC V8T 4P4 CANADA
phone 250 383 6864 (toll-free 1 888 232 4444)
fax 250 383 6804; email to orders@trafford.com
Book sales in Europe:
Trafford Publishing (UK) Limited, 9 Park End Street, 2nd Floor
Oxford, UK OX1 1HH UNITED KINGDOM
phone 44 (0)1865 722 113 (local rate 0845 230 9601)
facsimile 44 (0)1865 722 868; info.uk@trafford.com
Order online at:
trafford.com/06-0844

10 9 8 7 6 5 4 3 2 1

"However vast the darkness,
we must supply our own light."

~ Stanley Kubrick

2001: A Space Odyssey

To The Tao,

...the un-nameable something that created, develops, and sustains Cosmos – expressing itself through its infinite creations.

In this book, I admit, a darker side of things has carried over at times. I can not pretend I do not know what is going on here now. And so be it.

While the future string of events, on the horizon described, is slightly unsettling. I know I do my best to empower you. Times ahead will activate you even more.

You will free yourself from the known and transmit this further. You are part of it. Some will come and some will go; some will rise high, and some will sink low. Which it will be is based on a volitional decision among the enlightened ones. Intelligent Cosmos watches and guides all of us with ultimate care and love.

We the few who study *Zen of Stars* ask ourselves if we are on the right path. Know that we are not alone in this. We have friends out there – and in here. While dark ones use occult mind power, clans of light work with energy.

What is at stake is way too big for a messiah or savior to manage alone. In the age of knowledge we will make the quantum leap as a group. However, each one will do this rite of passage alone. Remember that we are the fine souls we have been waiting for.

This book shows – in non-linear ways – the leap of mankind. You are holding in your gifted hands a Cosmic time capsule. Forget The Past – Remember The Future!

Michael St.Clair

Song of The Seer

In this circle
O ye warriors
Lo, I tell you
Each his future.
All shall be
As I now reveal it
In this circle;
Hear ye!

Tatanka-Ptecila – Dakota Sioux

Which Future?

Zen of Stars presents potential scenarios of humankind's future. This book suggests that probable futures lie ahead of us in a state of sleep – or suspended animation – from which our actions and behavior select one scenario to make it our present experience. We manifest the future through our behavior.

While some of the predictions outlining future events may sound frightful, they are within the logical range of the extended now, or probability. Zen of Stars suggests that we have a choice to opt for, and create different outcomes. You are asked to keep this most sacred freedom in mind while enjoying and studying this unusual book of predictions.

In his book about decoding the lost science of prayer and prophecy, Gregg Braden explains The Isaiah Effect: Quantum physics and science suggest the existence of many possible futures for each moment of our lives. Each future lies in a state of rest until it is woken up by choices we make in the present.

St.Clair makes it clear that the future is not pre-destined, nor set in stone, but is selected from an array of different potential and possible options lying before us. We are the ones who have to select which future we are going to live, both individually as well as collectively.

The magician – one who creates consciously – delineates the science of how we choose the future we will experience. When we select a future, it becomes our active present. As St.Clair points out, our present is the future we have knowingly or unknowingly opted for, albeit often

9

influenced by fear or the uncalculated swing of emotions. We must move away from the unconscious play of emotions to see sustained magic happen.

If we wish a different future, we need to choose on a conscious level, a different present. The tools of our awareness are intent, imagination, intuition, dreams, visions, fluid intelligence and love. Above all else, the psychic navigator can help us to heal, as well as to choose and create a sane collective future.

What future do you want? Who do you want to be today or tomorrow? Where are you going with your choices? The direction you take and the responsibility are up to you. The key of St.Clair's message is "you decide". His record in predicting events is highly accurate, because he is able to see and understand the kind of decisions key people will make.

St.Clair stipulates that we can redefine the parameters of premonitions, prophecies, and predictions made over the last few thousand years. Nostradamus, Edgar Cayce and St.Clair's own astrological calculations are calibrated to help mankind avoid mass extinction over the next millennium.

Even if this book lists scenarios for a denouement of the world, St.Clair is not saying that it must be so. Quite to the contrary, he insists that we should opt for a different outcome. However, to carry on in the manner we have chosen so far, means St.Clair's psychic recall warning is the outcome to expect. There would be no reason to doubt these predictions based on the mathematics of time. His past predictions have already been proven correct.

Future star alignments could be used for other purposes if we so wish. *Zen of Stars* shows how meeting the challenges ahead can take us into outer space and to explore cosmic consciousness. St.Clair, more than anyone else, knows that we are capable of changing the outcome of premonitions and prophecies, precisely because we can predict them. If we become aware of probable outcomes we do not want, we open our consciousness to the possibility of creating other actions and choices with different results. This book is therefore about intelligent action and how we use the sacred freedom of human choice.

We came into this universe and onto this planet for one reason above all others: to give and receive a love greater than that known by ancient deities and spiritual masters. This love is greater because it will be a collective manifestation, individually and jointly chosen. A love that will transform the hearts and lives of each person on this planet, and thereby the earth we live on.

We were given the seeds of love and cosmic consciousness as our destiny. This cosmic consciousness is designed from its inception to further the evolution of the universe, planet Earth and its inhabitants. Yet, we on planet Earth were given freedom – sacred freedom – to choose. So the choice between evolving and self-destructing is really ours and not solely in the hands of God. In this sense, St.Clair's book represents a departure from conventional thought patterns.

Zen of Stars will make you question your belief system, reality and authority. It is an unusual document, a rare account, a fascinating yet thoroughly annotated study. St.Clair's work traveling into, through, and out of time is the culmination of a lifetimes work by its author.

St.Clair explains the mathematics of time, or why and how belief systems are our maps of reality as they shape what we perceive. He then goes deeply into the imagery and meaning of star icons and the hidden destiny of man.

What you are about to embark upon is the spiritual adventure of a historic process called space-time, or magic, *high magic*. It was said that high magic is the *Yoga of the West*, thus making St.Clair an occidental yogi. Magic is processed through the sixth sense, a study of which the magician shares toward the end of this book.

To foresee and predict the future as if it were already the past has been St.Clair's gift, vocation, profession, and occupation since childhood. Now, he lectures this procedure to a wider public. Colleagues have said he is the astrologer's astrologer. He has written *Zen of Stars* to make his knowledge available to the world, to promote sane choices, and teach: *The Way of the Magician.*

St.Clair explains the world's alchemical passages, its occult time lines, eternal cycles, ancient civilizations and modern cultures of war and of art. He discusses the world's perennial philosophy, archetypes, sacred geometry, and sacred centers. He shows how the past, present and – yes – the future of mankind is delineated.

St.Clair merges his own experiences and insights with his ability of psychic recall to foresee the times we are entering now. In this work, St.Clair decided to tell his own story in the form of a picturesque and imaginative literary thriller. This book is a magical work showing us where we stand now, revealing hidden times yet to come and helping to understand the choices ahead.

To be a magician is to assume one's proper place in the cosmos – by birthright. What that place is, each soul will discover on its own. By communing with the sun, moon, stars, planets, galaxies, and the spirit of cosmic consciousness, we realize the *One-ness* of all things. To become a magician, St.Clair suggests, is thus to become more than just human, since it is only in the true fulfillment of our maximum human potential that we realize and accomplish our destiny.

St.Clair's narrative amounts to a voyage through space and time, revealing his sensitive and vulnerable psychic existence. The fact that he is alive to further predict and implement the future he so accurately describes is the light of his own existence. *Zen of Stars* set against the backdrop of the collective memory of mankind – from Atlantis twelve millennia ago to well past the 22nd century of our time – is truly captivating.

St.Clair's passion about what he teaches reaches out through his presence to the clients he advises, and though lectures to astrologers and audiences. He demonstrates the magic act of faith in what can be done, as opposed to fate. It is an important moment in time that the magi-sage is willing to share with the world: The cosmic design of space-time and the evolving new society bound to emerge; a time where Pluto moves through the last four constellations of the Zodiac over the next half century, to herald our age of consciousness and to transform the world.

St.Clair draws pertinent parallels between recorded history, our present, the future – and major planetary alignments, which he has diligently studied and compared over many years of his life. In *Zen of Stars* he has finally brought together his theory presented here for the first time.

Zen of Stars is a witness of time, of the current Zeitgeist, in which St.Clair chronicles his meetings with men and women from the most diverse walks of life. In his work, you get to know the man of thought and of action, the magician who chooses and creates consciously and the man who manifests intentionally.

St.Clair's life shows that he has been in and part of, but not of this world, before he turned to contemplate it. He was in the beginning of his career trained to become a general, a lawyer and an investment banker. He then turned his vision towards communication and understanding, and to all things spiritual.

During his career, St.Clair has confidentially advised a number of dazzling international clients whom he helped to fulfill their destinies. He witnessed – if he was not instrumental in – the denouement of many private and professional trajectories. The personal web of fates in which the magician was, and still is a part of, is intimidating even to the most jaded.

St.Clair knows that you are the master of your destiny and he helps those who consult him to grasp this most important fact. Coming of age at a dangerous time, he is an advanced thinker and the planet's foremost astrologer. St.Clair sets the record straight on mysterious issues from Nostradamus' predictions, to Orwell's visions and to the Sanskrit mythology.

He walks us through the fulfillment of prophecy from Atlantis to the rise of the Aryan power, and the occult control behind it, to the earth changes and well beyond our time. He spent a good part of his life studying ancient texts and occult scriptures. Today he is essentially a sage, a magi of the Golden Age.

This is a comprehensive book about humanity's recall. It is not apocalyptic, but rather hopeful: it designs the new, tribal, high-tech, stone-age, and water-world type society without wars, borders, boundaries, blocks, or governments into which he predicts we are about to move after 2050. The gods will fill the universe is the future his awareness foresees with intent. He invites you to discover this for yourself, as his goal is to share the secret and sacred knowledge of the ancient star magicians.

This work situates St.Clair in the spiritual lineage of mankind's magic minds, as he represents their work, from Apollonius, to Nostradamus, to Meister Eckhart. Master St.Clair perpetuates a sacred tradition of thousands of years. He is at this time a powerful – and possibly the most accurate – magician alive. His astrological predictions stand the test of time, as you will see.

Unlike previous adepts and mystics who kept their knowing secret, he shares it with you intelligently. Turning this page, you enter a world of esoteric high magic, a universe of hermetic science, a realm of creative skills. You are initiated gently into an ancient craft, the holy ritual of spiritual astrology, the art of de-coding the galactic clock of destiny as space-time takes on an entirely new form and rediscovered meaning drafted by St.Clair the astrologer.

St.Clair lets you read parts of his own diary of magic, welcoming you into his blissful sanctuary of holistic knowing and cosmic understanding where science and mysticism merge into star magic. The gray zone – dawn of a new society – becomes vivid and colorful in this book about our collective human journey into the future and beyond.

Star magic is the art of cosmic creation. This master has created a work of art beyond time, pointing to the edge of nowhere. You hold in your hands an art book, an autobiography, as well as a brilliant legacy on predictions for the 21st century. In its subtle way, this beautiful achievement is also an anthology of the most supreme magicians, mystics, and scientific minds of humankind. Once every decade perhaps, one comes across a book that will make a major impact on the world. *Zen of Stars* is that book.

Argos

Advancing The Light

Master of Light

Naming A Star

Mirror of Seeing

Zen of Being

A Small Boat

Dawn of A New Age

Millennium Signature

Designer of Star Magic

Master of Light

"The world is changed.
I feel it in the water.
I feel it in the earth.
I smell it in the air.
Much that once was, is lost,
for none now live who remember it."

~ Galadriel, Lord of the Rings

The man from the future stood with his back to the wall of the grand dining hall in Chillon castle overlooking Lake Geneva. Wearing a navy suit, a black shirt and dark polished shoes, he looked the way a time lord would as he passed without motion through the corridors of time and into the space of my presence. Have you ever asked yourself: "What is real?"

The noon sunlight sparkled across Lac Léman and crystal reflections danced in his green eyes as a glint of golden stars passed through the Galactic Center.

Lines of bright light cut between the shadows across the marbled stone floor of the castle rock. It had dawned on me that during some thousands of years of extra-terrestrial interactions some very interesting souls had inhabited these walls in which I now peacefully stood, learning and applying the craft of seeing.

As I thought about this business of seeing, the Zen of knowing, I realized how this man was perfectly at home in the castle, which was after all his doorway between the worlds. He paused briefly and signalled across the bridge of time, as if he wanted to make sure that I would see him.

If I were to say in one word what I know he was or is then it is what he said I am: a *trans-human soul*. Radiant crescents of light rose to shape the shadows cast through the high windows of the castle. Misty mountains appeared across the lake mirrored in the Gothic windows, reminding me of the fact that they are the timeless guardians of this sacred place.

The impression I gained as I followed the elegantly dressed man up the stairwells to the watch tower, was that of a vortex of timelessness situated on top of a thousand foot high granite needle, emerging high above the surface of this mysterious lake that had kept so many secrets from this world. Was he me from a time yet to be?

The timeless one had come to teach me how to unlock and master time and how to see the invisible world for myself. He was definitely real in the three-dimensional sense of being, although I do not know how he came to stand there in front of me, on the threshold of reality and unreality. He was smiling this incandescent smile of his, and I knew something unusual was about to happen. The time portal opened as the eclipse passed its shadow over the earth, revealing the timeless mirror, showing me the things I had wanted to see.

He was – so he said – the master of time. He was here to impart to me the Zen of knowing and the art of seeing. Later, I would see that he was also the *Master of The Light*. He had come into this time as some sort of other-worldly visitor to talk to me about certain things I had wanted to know, such as the meaning of life, the future of planet earth, the origin of man and many other considerations of that nature.

What we would do about what he taught me was foremost on my mind. Everything seemed new to me, and yet I knew it to be true. His words resonated in my mind like déjà-vu. His mind became the bridge to a long forgotten knowing, and what I saw was not as I had expected. It was much more. The castle became a living portal of light and I saw beyond our earthly existence into a future too powerful to describe.

The heavy stone wall behind us shimmered slightly and opened into a beautiful place filled with light. In the space beyond the wall I watched colors dancing. Colors I had never seen before. Colors we have no names for. It really is not necessary to name everything we see. It is more important that we just take it in and behold it. This constant need for naming things is not sane, nor is it helpful to the deeper understanding of things.

People from another futuristic looking time seemed to be at work in this space beyond the walls of our reality. What they were actually doing and working on was not so clear to me, until much later, but everyone seemed to be at peace and focused on something that all of them considered important. They were finding *Nirvana* while experimenting on some complex time loop mechanism.

The people I observed in that window of timelessness were Zen masters of sort; they were both spiritually and technologically highly advanced. Like a gentle rain, sparkling rays of rainbow-colored light flooded into the vortex chamber and merged with the master of time standing close to the wall. He had opened the mirror of seeing to me. All I had to do now was to look into it and see for myself.

21

How sad that most people alive today have forgotten the rainbow people. Only my friend, *The Master of Light*, has not forgotten these things. His language was his silence. His action was his knowing.

He turned to look at the light cast through the mountain time portal – a window high above the lake. Tiny rainbows of light moved with precision across his forehead merging with his third eye. A perfectly round crystal floating in the doorway was the mirror fragmenting the seer eclipse. The same rainbow colors appear on dark shells growing on the rocks beside the seashore. If you look closely enough you can see the rainbow colors of light within the dark pearl of the oyster shells.

I was told that the shellfish listen to the sound of the moon passing slowly over the face of the sun, and it seems as if the clans of light also listen to the message of the planets as they move into their clockwork-like positions, silently but methodically, timeless and eternal.

We do not only function in one time line. We are of the future meeting the past in the now. We exist already in the multi-verse of uni-verses, in realms we do not yet know of that exist to assist the realm we think we live in.

The portal of light, shaped as a phi-spiral vortex, closed as silently as it had opened and the space beyond time once more merged with the heavy stone structure sitting on the giant granite needle by the lake. The seven high pillars holding the vortex chamber in place returned to their original form – known today as the dungeon of Chillon – where Lord Byron had scribbled his autograph two centuries ago.

The man from the future walking back across time

seemed to shimmer with the light of the portal wall. Without his wisdom the rainbow tribes would be forgotten forever, and human beings of this planet would lose their connection to the cosmic beings of light across the bridge of time.

We were back in my own time, the 21st Century of our modern world. I had just seen the future from the past. It was clear to me then that linear time was an illusion.

The master had crossed from another realm, some higher dimension, one we know as the future, and into our time when Pluto in Sagittarius, its ruler Jupiter in Scorpio, Mars in Taurus, Neptune in Aquarius and Saturn in Leo had formed the grand cross. This man began to answer my unspoken questions in a wordless dialogue. Was he myself looking in from a time yet to be – via some bi-location, like in the great novels of science-fiction? Is it possible that the future was meeting the past now?

Wordless, did I say? It was a change of worlds for both of us. Telepathy, a form of mind to mind transmission, continued effortlessly like a silent holographic download on a computer hard drive – his medium of choice – before he disappeared into the vortex of light... He passed silently through the castle walls, out across the lake and into the sunset while a defragmentation of knowledge was taking place inside my own being.

The planet Venus, high above the misty mountains, glistened golden in the dark purple evening sky as I recalled what we had exchanged in one day: lifetimes of magic, the Zen of Astrology, compressed into one sunset. Who wants to live forever? That question remained unanswered, hanging in mid air, as I left the castle. Walking across the moat and out onto the old cobble stone road of the 12th Century to return into my own time, I knew that the here and now lasts forever.

Overlooking my lake from high above, beyond my mountains, I can look the truth into its face unfazed. Truth is the new, and rarely the known. My lake, formed like a large crescent set like a diamond in the mountains, is shaped and behaves in magical ways – almost always still, like a mirror, reflecting the fire from the sky. In fact it acts as my mirror of seeing. It recalls timelessness and simply sees. It does not judge what it sees. It just sees and mirrors its recall as a witness, like a living timeless reflection. The lake showed me that we are all seeing mirrors once we remain still. It invites me to look both deeper and beyond.

What was it again that *The Master of Light* and I spoke of together during our conversation, in our meeting of the minds, in the presence of the invisible world? He had said to me that *time is the root* – and that we begin here now, outside the man-made time-warp. He also said we would meet again. He would introduce me to his friends. Now I knew I was not alone. I also had friends in this and the other life. With a profound look of calm knowing, he had beamed this book and the following thought at me:

"You wish to run away and hide, but where will you run to? Your own SELF is where you LIVE. You cannot escape it. Stay with – and develop – it!"

Naming A Star

Frodo: "I wish the ring had never come to me. I wish none of this had happened."

Gandalf: "So do I. So do all who live to see such times, but that is not for them to decide. All we have to decide is what to do with the time that is given to us."

~ J.R.R. Tolkien, Lord of the Rings

We tend to observe least what we know most. When I look at the lake, or at a star in the sky, I do not have to know its name. I just see and behold it. I sense that in order to see the star I really would like to touch or experience it, minus the concept of knowing that it is a star, and not to know it so much as such. Naming it would only put a word between me and the star. Giving the star its name puts a veil of sorts over it, instead of building a silent relationship with it.

To name Antares the heart of the Scorpion, or Aldebaran the watcher in the East near Orion, the hunter... or whatever else we know about stars – is not **Seeing** the star, but it is merely thinking about it, by naming and comparing it.

Comparing things is, in my opinion, an act of mental violence when we care to think about this deeply. The beauty of it – the star – is in itself and me, or itself and you, and not in the comparing, or numbering, tracking and naming it.

I found out when I was a very little boy, some forty years ago that I could look at the night sky and the sky became magnified. This seeing was natural, it was an action of its own, without thought. I really could not have cared less what a certain star was called. I did not even need to know that it was called *a star*. Later I would find out that it was not even a set thing in the first place. Star or not star became an interesting pastime as time passed by. This is how I came to be the *star magi*, by seeing as a child sees. I knew that I had come from the stars before this life here even began for me, and that seeing is what I was born to do. Seeing brings its own intelligent action.

One night, not so long ago, I was looking at the stars with friends. We were looking at the crystal clear midnight blue sky with intent. Now granted, as an astrologer I am supposed to know the heavens and its stars, and I do – to a certain extent. That night, however, I was just looking, or seeing, while people were chatting close by. Then, all of a sudden, a brilliant white light lit up due West. I smiled and I pointed the light out to my friends. This was not a normal star. They saw it too. It was there as if from a command of my mind. I magically willed it into being, so we could share the seeing together.

Later it occurred to me that perhaps the light sent its thought into my mind, so that I would look into the direction where it would light up. This is the kind of game a star magi plays. It is during those types of mind games in playful times that I shift to think about things of importance.

It was clear that this light was not a star. I knew this because where we were looking there should be no star. It was not a planet either. The planets had set shortly after sunset. It was not an airplane either. The solitary light stood still for a long time, perhaps over an hour at the least, blinking and flashing multi-colored lights from afar. My Chillon Castle, the time portal, was close by too. There was time enough to see it, to look at it, and to *think* about it. Things of least importance were being moved out of the way of things of utmost importance.

The observer, the observed, the castle, the lake, the night air and the mysterious bright light in the sky resonated with one awareness. Without going into what this light was or what it was not, I was sensing that maybe the light was seeing me and my friends. It seemed to be signalling. The observer now was the observed, and I found that very funny because it reminded me of the sage and teacher Krishnamurti and of his views on consciousness and the ending of time. All of this happened after I had explained to my friends what the shift of the dimensions would be about and how we would eventually *get it* and live with a new awareness.

I thought I would share this impression with you, because it will set the stage for what I now want to explain. The word, the name of the star, will never help us to touch the star, because the naming of the star, which is merely astronomical knowledge has conditioned some of us so heavily that only the name has become important, coming between us and actually seeing what we see. The same goes with naming animals, plants, and human beings.

It would be wiser not to name things and instead just *see* them, just for the sake of observing them. It occurred to me that the one who sees and the object that he or she sees are actually not separate once we drop the conditioning of our petty minds and feel the newness of each moment. Let us keep feeling that we are new. I do not know what the light was that I saw? It is quite all right not to know. Sometimes not knowing is its own protection.

During that night by the shores of Lake Geneva, with the light watching me, my friends and my lake, the seeing made an impression utterly clear to me. Knowing is its own right action. It was not the first time that a mysterious light had appeared in the skies when I thought of one. The Nordic Blond ETs had told me via telepathy to look for one and to see it mirror itself into the silent lake. That was when certain things became quite evident to me.

As The Master of Light had suggested:
"Stay with it!"

Mirror of Seeing

"The person who takes the banal and ordinary and illuminates it in a new way can terrify. We do not want our ideas changed. We feel threatened by such demands. 'I already know the important things!' we say. Then Changer comes and throws our old ideas away."

~ The Zensufi Master
Chapterhouse: Dune

I was called the *Obsidian Mirror,* or *the presence with smoky vibes that sees.* My star wisdom is the magical knowledge of what to reveal and what to conceal, and when and to whom. The time lord decided to make available to you what we know you can handle, which is more than you think. We begin here at the ground zero of seeing.

The lake of life is its own mirror of seeing and of being. We begin this journey – the adventure of Zen magic, journeying into the presence of the unknown as if we knew nothing. This will make our life easier, defining a new beginning after a stagnant age of ignorance. We will make this trip with no fear.

Fear is based on time. As an astrologer I know time simply transforms things and people in different ways. For me, Cosmos moves both rather slowly and yet at the same time very quickly. Agonizingly slow it seems because I see things as they are, while at the same time almost everything and everyone floats by at warp speed and super-fast. I do not like to count the planetary revolutions for myself, as these passing seasons are to me ripples, forever repeated in an eternal cyclical and almost circular stream of planetary dancers. And yet beneath the central sun of our galaxy this rich pageant called mankind will find its end at some point in time – if I am not totally mistaken.

Many worlds and planets have suffered through multiple traumas, passing over eons enduring unspeakable troubles, wars and cataclysms; and yet the past is past and the trauma is now over. There is no point in remembering and analyzing the collective trauma. Trapping us as the regressive forces would like us to do. We have no choice but to end the suffering now, it is that simple. It will only end, if and when we, you and I, *end it*. The decision is yours and mine.

This change begins at a personal level from psychological to psychic, and goes on all the way to the inter-dimensional levels. There is no point in analyzing and feeding psychological trauma. We must end the past and not nurture it through thought, with the result of keeping it going. Therein lies the trap, to keep looking at past events over and over again, from wars on Mars, to the ancient Mayans, to the Cathars and Celts, to Tibet and more, to the last wars of the 20th Century... End it!

We can begin anew, at any moment in time. As for me, that moment is now. We have to realize we can do it. Healing the heritage is an idea, and remains an idea, unless we actually do it. As I wrote, the things of least importance have to get out the way of the things of utmost importance. What I am talking about here is *not* some intellectual thought process type of stuff and I do not wish

30

you to *think* it through. This change is real, cosmic, planetary, individual, instant and complete as it is. People think a lot and like to think things over again, instead of taking direct action without thought. For now, I invite you to see. Let me share this excerpt with you from Lord of The Rings, Galadriel says:

"Here is the mirror of Galadriel. I have brought you here, so that you may look in it, if you will."

"What shall we look for, and what shall we see?" asked Frodo...

"Many things I can command the mirror to reveal," Galadriel answered...

"And to some I can show what they desire to see. But The Mirror will also show things unbidden, and those are often stranger and more profitable than things which we wish to behold. What you will see if you leave the Mirror free to work, I cannot tell."

"For it shows things that were, and things that are, the things that yet may be. But which it is that he sees, even the wisest cannot always tell."

Do you wish to look? I assume that since you hold this book in your hands... you do wish to look. Some of us human beings, the seers, shamans and sages among us, have refined and excellent visions about how this world would best look and should be, while the beings watching us, and the presence I call *the force* that works with some of us, laugh at us, benevolently.

We keep dreaming and expressing ourselves about wonderful concepts and strategies on how we could improve this world, hoping or wishing that by doing so we will *auto-magically* create motivation leading to actions for certain individuals and groups. All this is great, and has been tried before and... evidently did not produce the desired result. Why do you think this is so?

To be an individual means, by its own definition, to be undivided or non-divisible, and yet are we whole as individuals? Not really when you care to look at it very carefully. We divide soul from mind and mind from body, love from work and action from thought. Therefore we are actually not as *individual* as we think we are.

However, a few of us are starting out on an entirely different approach: an action without thought, or what is called *direct action*, in which concepts such as ascension or living without ego are set aside. Change begins only, at least in my opinion, when the old has died. Only then can healing the heritage and forging the future begin.

What I am saying, is that for us to become who we are supposed to be, we have to actually die to our old human self-image, and let go to transmute into what human beings are in their purest form.

This world, planet Earth, is full of unimagined potential and possibilities. It is the home into which we are sent when we separate from our original source. To journey, for a moment, in the illusion of separation. But we are actually not separated. We have invented all sorts of concepts and thought structures which are nothing but ridiculous illusions and excuses for not doing what we are here to do. The fact that this earth has hosted so much conflict in so many thousands of years leads us to assume that nothing can grow here anymore. This assumption is a mistake.

We presume that because we seem to know there will be some more conflict the light will withdraw; when in fact the light is actually getting so immensely bright day by day that it is separating the dark ones away from the light ones. The light I am talking about – the one I was shown by the *Master of The Light* – is cornering the darkness of this world. No weather manipulations, no wars of any kind, no domination tricks, nor mind control methods by the dark ones will overcome what I know as *the force*.

Is it possible that we can enter into a new world where we can solve the problems caused by humanity? It is indeed possible that creative dialogue and thinking together can take place even between the oldest rivals of all dimensions and nations and groups, if they die to their old selves and create their own newness. The light beings looking in from outside are *laughing*, or at least shaking their heads when they see human beings suffering. We did not come here to suffer. That too is an illusory thought construct.

The freewill above self-predestination we were given is known to me as direct action – not a choice mission with its opposite anti choice. It is the freedom of pure action that we are given as ambassadors of creation. If we are indeed to have been made in *God's* image, and if we assume that whoever God is created this creation, then we would logically come to the conclusion that we too are here to create.

Who created the creator? Creation is direct action. It is an action that works together with a co-creative element: Cosmos and its forces. This is what I know as the *intelligent cosmos*, or the space in and around you and me that communicates with us at all times. The information I share is based an ongoing dialogue I have with the intelligent cosmos via the *Master of The Light*. In a form of noesis, or guidance, the intelligent cosmos uses a power by which it steers those of us who are able to blend into this intelligent cosmic power in the moment.

It does this by sending us signs, via animals, human beings, events, situations, circumstances. Events some would call serendipity, synchronicity, noesis or whatever you like to call it. I call it the *seeing mirror of Zen magic*. I was asked what Zen magic is. It is the art of making things happen, it is creation by merging with the Presence.

What I know as *the force* has no counter force. Some sages call it love. Love or the force, is ultimately intelligence and is self-organizing, as is Cosmos. It is what moves Cosmos. It will at times remind us of what we were initially given via creation. This force is all-creative and uses guidance as its messenger. An inner guidance, whose origin is yet to make itself known to us. Once we get to recognize and make friends with the force, we can move inside the force field and merge as one.

In the moment we are one with the force, all concepts, ideas, and strategies we carry around become secondary. Then the actual – not imagined – but actual creative action begins. It is at this point that change manifests in our lives. Then direct action leads the way. To come to that point of realization some of us have to traverse an intense passage, the dying of the old self and becoming as Cosmic as a planet, a mountain, the ocean, a sea, or a river.

When you follow any river to its source, you will find that it begins with a small drop of water coming out of some cave or glacier on top of some mountain. A tiny drop from somewhere up in the Himalaya will become the

Ganges river. A similar drop from a glacier in the Swiss Alps will become Lake Geneva, or the Rhone, ending in the Mediterranean. We the few, who have grasped this, are that first drop of water and therefore we become the ocean of change.

On the day I met the *Time Lord* from the other dimension; I was told by the force: "You Are the Mountain!" And the mountain I like to go visit and be with has let me know in its own ways – via the force – that at least a few of us are to be like the mountains walking the earth. Or like that ocean or planet, with no self, but with the capacity to produce action which will induce change.

This form of "death" I witnessed is more devastating than the physical death people are afraid of, as it will lead to a totally new way of doing things. In other words, to me drama and trauma is no longer my world, so what will my world be? Short answer: *Living inside the mirror of seeing*, the *Mirror of Galadriel.* As you will see, it is easier to die and go live *in heaven* than it is to radically change the mindset in the way the sages have suggested throughout the ages. To begin our own newness in this very life right now, while still alive, is an ongoing process that will never leave us. It is, I trust, what teacher Jiddu Krishnamurti once wrote about and called *"The Path"*, which is the pathless path.

As humans re-incarnate on earth, then surely we would be super-smart and all-wise by now. So why is it that so many of us have died in other lives... and yet we are still behaving as we did eons ago? This is what I mean. This is why the death of ego in one's own real life now is more important than the physical death so many of us are afraid of.

The few, who have seen the force in themselves, are now learning how to work with this force and how to handle it constructively and for the good of mankind. We know that the choices we make are inherently guided by our intelligence working together with the force.

Via light beings, guided *ETs*, and certain human beings, the force sent to earth some *special ones* to elevate humanity to its proper level and to teach mankind how to heal from the *trauma* of being human. Apparently most human beings are not able to access this power from the other world. Many sages and shamans, teachers and healers were sent over many millennia to help man change.

A few will receive something from this impression the time lord shared with me. Truth is never the known, it will always be new. I cannot speak of the new – of this amazing change in process now – in the dying language of the old brain. And so, in this book of Zen magic I use the language of the *seeing mirror*. Only the few who are prepared to step into their own power – via direct action – and move with the force of the intelligent cosmos will be able to make the difference. This difference will be made by *applying* the power given to us. It means we are the action which will bring change, which means to change to an action without thought.

If one human being makes the correct choice, then this one human being will change the course of the possible future in very relevant ways. No human being should ever feel meaningless or insignificant, because it only takes one human being to alter the consciousness of humankind through the *Cosmic Force* motion that creates and shapes all of life. One human thought influences another thought, then another, and another, until eventually this initial human thought manifests reality throughout all of Cosmic creation.

It is the same human thought, and the same human life force that causes an entire group to change course, as the group acts with one undivided mind. This oneness transforms the fragmented illusion of separation. As planets chart their prescribed courses and form their myriads of aspects to one another, so do we human beings engineer our own destiny, by using calm, determined, and focused energy, by using intent and being of one Cosmic mind.

As Neptune and Uranus receive each other in their mutual signs they tend to bring out the best in each other until 2010, and the one Cosmic mind is one manifestation we will see applied soon. Beyond that point, 2010, we will move toward blank slate technology and the defense of planet Earth.

You may think you are only one human being and thus unable to effect lasting change. This is not so. Look at a few names I picked at random to witness who brought about radical change of mind with their one single mind: Nikola Tesla, Nicolas Copernicus, Michel Montaigne, Galileo Galilei, Wilhelm Reich, Immanuel Velikovsky, Jiddu Krishnamurti, and HH The 14th Dalai Lama.

These are people who simply never gave up trusting in their vision. Who never gave up living their vision. It's never too late to change the way you live and at the same time change the world you live in.

Zen of Being

"If the feet of enlightenment moved,
the great ocean would overflow;
If that head bowed,
it would look down upon the heavens.
Such a body has no place to rest.
Let another continue this poem."

~ Zen Koan

Born with Saturn in Capricorn in my first house, I am both creator of chaos and organizer of order. You enter my space and presence by leaving behind all pre-conceived ideas and knowledgeable notions people have put in your mind. You leave me, after your visit, as an informed entity. To be an informed entity you must stay non-aligned to any conditioning of the mind. You question everything – even what is written in this book.

I invite you to dispose of all psychological and thought based obstacles to clear understanding. This is the advantage of the adventure called *Zen magic*. My world is form and formless... ever morphing into its new beginnings. I know that my power is self-organizing and self-contained in the new. Your willingness and fondness to leave behind old paradigms will help you to attract the unexplained.

In a way I am a prism that captures and throws out light beams in countless and unknown colors, and into many different surfaces and shapes, some of which are human by design. It is up to you to catch some rays of knowing into your own essence, transmute what you can use, and transmit what you deem fit to send out and share with others. *"We are cups, constantly and quietly being filled. The trick is, knowing how to tip ourselves over and let the beautiful stuff out."* – Ray Bradbury

The smoke-screen of the Zen master functions as a realm of I-Magi-Nation, and it does not have to separate Saturnian reality from *Neptunian surreality*. Who is to say what is true or false, and who is to know what is real or unreal, since this vast universe cannot actually be grasped by a single sight, sound, smell or any other known means of discernment?

Does the etheric world have an astronomical locality? Probably I would say yes to this question, but I would be hard-pressed to point you to its location. Sir Oliver Lodge and Sir William Crookes were pioneers of radio and television, and as such they were able to put forward a rational explanation to account for the seemingly super-natural phenomena appearing at their experiments. They said that this etheric world is the same place as our radio and television signals, but at a much higher frequency. Recent discoveries in quantum mechanics, the study of the building blocks within the atom, completely vindicate what these great physicists said at the beginning of the century.

They were adamant that we all survive the death of our physical bodies. That death is a mere change of worlds. The experiments by Lodge & Crookes proved that we have at least two bodies: one finite body containing the brain that dies, and another in-finite body, which plays host to the *mind* that separates when our short stay on earth is over. Lodge said the beings from the next world appearing at the experiments must possess bodies that are made of the same invisible matter similar to radio and television signals. They called it an etheric substance. Today, other equally serious scientists and searchers suggest that our universe is electric. I say Cosmos is an infinite intelligence beyond any theory invented by man.

Long before current science tried to describe the mysteries surrounding us the ancient cultures had long held the view that the physical universe is formed out of the invisible world, through *ether*, as it was known then. First there is the world of the spirit and out of this the world of physical matter forms. The invisible world gives birth to physical reality.

The Celts, Tibetans, Toltecs, Mayans, Native Americans, Aborigines, Hindus and most original humans on this planet all knew this most basic truth. Their way of life was based on this knowledge. They interacted and learned from the world of Spirit and sought its guidance. Will it take the next 13,000 years to rediscover what was once known?

I was six or seven years old when I began to tell people about the things I was being shown by the invisible world, the beings I called *my friends*. They were the luminous beings I could see by day or by night and they were indeed friendly to me, whereas humans appeared rather hostile, as seen from my childlike perspective. Today I know I was not the only one who had these types of experiences. For me, seeing the invisible world was normal and natural.

Some of my family said I was a changeling, when they wanted to tease me. What they did maybe not really understand was that the joke would soon be on them. People told me that the high elves, the invisible ones, had taken the human child and left one of their own in his place. I guess that is why I have always felt I am not really of this world albeit in it. This would explain why I felt somewhat out of place here, surrounded by people whom I deemed a little bit strange. I was born many weeks premature, but then this was before I was perhaps exchanged for the other one.

The low level psychic recall abilities and skewed mental development, together with the stone-wall denial of the invisible world was something I could not figure out nor understand. But what was much worse was this: Most would not pay attention to me when I talked about the things I was seeing. I learned therefore to keep most of my findings to myself.

I was working on the mistaken assumption that if I pointed out the existence of my luminous friends, then people would look and see. After all, how could they miss seeing what was standing right in front of them? If I saw this other reality why could others not see it too? It was the same for star ships and the extra-terrestrial lights in the night sky.

In 1963, shortly after JFK was murdered, I was nearing my 5th birthday and I distinctly remember that my journey in high magic was starting at that time. One thing that bothered me deeply when I was a kiddo was the fact that I could not cope well with the inherent and self perpetuating ignorance of this world. At an early age I was, by nature, curious to understand important things about life (which is normal when born with Saturn in Capricorn in the first house). The other kids were content with playing ball games and dutifully doing, what was to me, their asinine homework.

Soon enough, aged 12, I was seriously studying chess, and then at age 19, after I got bored and realized I would not make it to become the world chess champion, I immersed myself in the Zen of astrology, or the mapping of the complex sacred geometry of birth charts. Meanwhile I still had to study law and political sciences to survive in the real world.

I sailed through being aggressed by family and friends, by teachers, colleagues, and later society as a whole when I would discuss some of my views. I escaped at age 22 into the Swiss army and was on my way to become a professional instruction officer when I figured out that maybe I had something more interesting I could do. Many years followed in which I did many exciting things, met many interesting faces in many exotic places, but still I remained who I am, in another world of my own.

It became evident that while my luminous friends were looking after me, there was also a rather dark presence intent on stopping me from divulging what I knew. The dark force was and is unseen, non-physical, yet it does cast visible shadows across our lives, often in form of misguided human beings. The darkness is its own enemy.

The things the invisible world of *noetic guidance* had shown me all turned out to be true. Each teaching, and instruction, each warning and admonition became an obvious and undeniable fact as daily life progressed over some 40 years from 1965 to 2005. Would people then listen? Not really. Only a few smart ones saw the truth for themselves.

The ultimate denial of the truth was the standard accusation: "You brought it about, you made it happen." I then understood what this reality was all about, and from then on I began to tackle my existence and life in general in a distinctly different manner. It was as if the very act of warning people to change their ways in order to avoid the inevitable consequences of their actions was in itself an act of malicious intent on my part.

So what is this deeply rooted denial about? Few people want to take responsibility for the life and reality they create. Life is a strange affair. Humans are living in a state of world-wide amnesia, moving away from truth and its consequences. Human beings have come from the stars. To continue to deny this truth is to give in to the implanted fears designed to imprison us in this 3D illusion.

Mankind has come from the stars and the star seeds among us awakening to this truth are here as teachers, sages and healers. This is a fact many will come to realize in the next few years ahead, and will put an end to the wars, conflicts and violent divisions we are facing in this present time.

Before we go into understanding the deeper mechanism of denial about the origin of man, and before we look at the very mechanism that a small number of us have deleted from the hard drive of our DNA, it is important to establish an inner form of magic, or Zen. I have coined this unified action: The Zen of Magic.

What is Magic? The art of creating change. And what is Zen? Seeing into one's own nature. The Zen of being, to me, is the seeing of magic *applied*.

The luminous beings from the invisible world suggested to me, for the purpose of my own quest, to get to know The Zen of magic, which involves the highest levels of inner knowing. It is their gift to me and to you, a DNA inherent gift each child receives at birth. In simple terms it is the human birthright, or individual sovereignty. Better to use the gift now than lose it.

Seeing is akin to Sagittarius. Knowing is akin to Aquarius. This means that people born with prominent Aquarian and Sagittarian positions should not lose sight of these inherent skills. Using the skills you are born with cancels out the world wide amnesia and the cultural denial, once you start looking inside yourself instead of studying so-called authorities. You see when you want to see. You obtain knowledge when you wish to know.

The invisible world did not show me how to deal with the denial of other people, since that is not my business. They taught me to deal with my own state of mind and with my own state of seeing. Once we master the inner reality, the other falls into place.

In a simple but powerful way, the teachings of the luminous beings of the invisible world show how the inner world of the human being is the key to our existence, and why it is indeed the portal to knowing and the doorway to being at peace. As a first step to this new state of mind into Zen magic, you learn to be sovereign, which means to be answerable to no one but to your own self. You learn to retain a peaceful inner knowing, an inner certainty about the things you see and the multiple realities you are aware of.

Knowing that strangeness co-exists with weirdness and that we are as strange to other worlds as they are to us, this is basically step number one before we enter the pathless land on this elusive journey towards cosmic truth. The Zen of magic is our own inner life garden, with our own inner landscape. We are the architects and builders of this fascinating, rich and varied world we inhabit. We are the designers and farmers of the pathless land. We are its creators as well as its inhabitants.

This is why I told numerous politicians and two world leaders that we need no constitution to tell us about sovereignty. Either we are or we are not sovereign. No piece of paper can tell us this. And on my way to Zen magic I did study the human and cosmic laws, i.e. the low level law that is taught at a university law school, as well as the high level Cosmic Law that only the cosmologist astrologer can access and decipher.

In the process of looking within for answers, or for inspired solutions to the outer world, we may one day discover a wise Zen monk – maybe hidden in our own self. He or she might look like *The Master of Light* I introduced you to at the beginning of this book. If this master says something others do not want to hear, or suggests things that upset people along *the way*, the best we can do is to listen and answer: *"Please forgive me, I did not mean to offend you. I am certain you know better. Of course you are right!"*

Always leave space in your plans for the universal order of spirit to manifest. We should plan with intelligence and listen with intent. As Zen magicians we can predict what the future will bring if the present events run their full course. Yet we can also change a potential outcome by transforming our thoughts and actions to effect a different future, manifesting the probable in the here and now.

On Sunday, 20 December 2020, and during the solstice of that moment in time, Jupiter the wise and Saturn the bringer of form move into and meet in the first degree of

Aquarius, highlighted by the Sabian symbol: *"An Old Adobe Mission In California."* This talks to us about the necessity for overall orientation and an underlying stability in our personal existence into that zone of time. Yet the psychological climate then will be transpersonal, highly networked and moving towards trans-human. The static symbol of the Jesuits outpost in the western world meets the eastern Zen on the shore of an ocean of infinity. There is durability in this moment when 2021 opens a visionary era.

A Small Boat

"He had bought a large map representing the sea,
Without the least vestige of land:
And the crew were much pleased
when they found it to be
A map they could all understand."

~ Lewis Carroll,
The Hunting of the Snark

Thich Nhat Hanh, Vietnamese Zen master and Buddhist monk and the Founder of Plum Village in France wrote:

"Many of us worry about the situation of the world. We don't know when the bombs will explode. We feel that we are on the edge of time. As individuals, we feel helpless, despairing. The situation is so dangerous, injustice is so widespread, the danger is close. In this kind of a situation, if we panic, things will only become worse. We need to remain calm, to see clearly. Meditation is to be aware, and to try to help.

"I like to use the example of a small boat crossing the Gulf of Siam. In Vietnam, there are many people, called boat people, who leave the country in small boats. Often the boats are caught in rough seas or storms, the people may panic, and boats may sink. But if even one person aboard can remain calm, lucid, knowing what to do and what not to do, he or she can help the boat survive. His or her expression – face, voice – communicates clarity and calmness, and people have trust in that person. They will listen to what he or she says. One such person can save the lives of many.

"Our world is something like a small boat. Compared with the cosmos, our planet is a very small boat. We are about to panic because our situation is no better than the situation of the small boat in the sea. You know that we have more than 50,000 nuclear weapons. Humankind has become a very dangerous species. We need people who can sit still and be able to smile, who can walk peacefully. We need people like that in order to save us."

I have studied much of the work by this Zen master, the work of an extra-ordinary man. We need people who have the knowledge, who can sit still, who can smile and walk peacefully; even though the dangers faced are very real.

Today most of the people of the world are asleep. They have forgotten who they are. At the same time a greater plan, as yet unseen, remains hidden in the chaos. This plan is unfolding and will soon emerge whether people are ready or not. The present day enemies of the earth's people will one day turn around and ask us for help. They will ask us for help because of the dramatic change we are about to make. The people on earth are destined to walk out of the war of violence cycle, and become healers of the cycle of love.

Our enemies are maybe stronger now, but one day they will be weak and frightened. The galactic energies are changing, and the DNA will become unstable in atmospheres of violence and hate. Our enemies will begin dying and will need to find a solution other than conquest and war. They will know why we have come here, and they will be ashamed to see themselves in our light. We will deal with them at that point, from a position of truth.

The changes about to happen until 2012-2014 are daunting, and yet humanity will make it. The passage of Pisces Neptune will be odd. However, we look forward to peace in a few years or decades. Thus, my overall message is inherently positive. How we will reach the other side of this *passage* and what issues we deal with in reality are the subjects of this book. The transition has already begun and will continue until 2021, when Jupiter aligns with Saturn in Aquarius.

I have learned to question everything, and in the process, I have discovered it takes great courage to come to a foreign land to find truth. The foreign land can be another continent, another way of life, or another dimension, depending on whether you are a time lord or not. Beliefs are not going to help us in this quest for peace, but sacred knowledge will surely be useful.

The key to the future lies in the past. You once had the knowledge I will share here with you, but you somehow managed to lose it. Let us regain it together. We are living in interesting times. Our path lies before us as luminous and full of Cosmic light and love as we wish to co-create it now, with the help of Cosmos – our *Creator*. In essence we are all Cosmos, Greek for *order*.

In the surrounding world, here on this planet Earth, the insanity is global and pervasive, not just peripheral. It is all around us, with and within us. We must leave the system; isolate our sanity from the dark forces and the negative alien energies, to set up independent and spiritually functional smaller human communities in order to re-learn the truth.

I am giving here an overview, a bigger picture of what

is about to happen and what we can do about it, as I present the true future in light of the true past, as well as practical solutions for living through the uncertain times ahead. But do not focus so much on what I have to say – instead concentrate on your own knowing as you read. Find out who you are, and your purpose for being here. This is my message: "Get to know your own self. Go beyond what I say!" The solutions are within your own mind.

The world has gone mad. The earth is under siege. Conflict is everywhere. The crisis can no longer be denied. So, to begin with, do not deny it. Do not try to run away. Accept the fact and deal with it!

The lies and tricks of the "age of deception" are monstrous and monumental: 9/11, the wars in Middle East, the hidden dark ones and more to come. The so-called *axis of evil* is the conglomerate known as the secret leadership of the dark ones, and human ignorance is the axis of insanity, which is more to the point. This cabal with its secrecy and greed has enslaved the world to its aims. You contribute to their agenda if you participate in the system, even if only passively by existing within its structure and tacitly acquiescing.

I like to look at the stars at night because they tell me they are there for us. They just are – and they are not at war. You can see the delusion we live under when you look at other humans or stare out of the window, watching how we all live, the insanity of the media. You can feel it when you go to school or work, attend a business meeting or church, or when you talk to your friends. Wherever you turn, you observe rampant insanity and dysfunctional human behavior. The fact is this behavior is actually no longer human. We have become the aliens of earth. War or peace? Most people are at war.

The great sages, scientists and seers from deep space contact times have long ago agreed on one issue: "Humans are asleep." Humankind moves through life unaware and in an unawakened state. Humanity exists and lives in a fallen world, among the fallen ones.

I am certain you have had dreams that felt real, more real that this world you live in. I have experienced such dreams many times and through many lifetimes. It is not new to me, so why would it be new for you? In fact, I will ask you this: Do you know for sure where you were last night while you were asleep, dreaming? Perhaps you do not know. Perhaps you cannot answer this.

Do you know for sure what you did yesterday or today while you thought you were awake? In what way is your waking world different from your dreams? How would you know that you really awoke from the dream? How would you know the difference between what is real and what is a dream? Is the dream real or is reality a dream?

How do you define reality as you live your daily life? Are quantum realities real? Or is reality multi-dimensional, the strata of dream-worlds, or all of it together? Are you living close to the ways of Cosmos, or are you trying to force through your own will in order to control and manipulate petty outcomes in your favor? You can believe whatever you want to in this life. That is your prerogative, and is yours to decide. Reality will meet you only when you are able to observe it. When you see into the mirror of the soul, reality changes in quantum ways.

The system we are living in is about control, greed and secrecy. It is about having more than we need (at the expense of others), and it is about hiding knowledge from others. The good news is that it is just that: a system, about to crash. Most people are not ready to let go of the current dysfunctional system. A system built and maintained by those who believe in its rules. The forces of decay are hopelessly dependent upon it and will fight to keep it. When it finally falls apart, which will be soon, then we will have peace.

The habitual mind does not adapt easily; it experiences discomfort in letting go. This habitual mind created the system ruling the world at present. As long as the fear-based system exists, the human race will continue to be in bondage. Is it possible for an intelligent human being to bypass it and let it implode on its own? I will show you why and how it will implode and end the illusion of the need to fight it, or bring it to an end.

Reading Richard Bach's book *Illusions, Confessions of a Reluctant Messiah*, I learned that one can take any book, ask a question and open the book wherever you choose. What your eyes light upon is either what you most need to read, learn or is the answer to your question.

In the next chapters – which are the equivalent of the lost footage of a movie left on the cutting room floors – you may find certain things you may not feel too familiar with, and yet you may also feel that what I describe is true. The issues we face are known to you, so I am not here to tell you how to live. I am just trying to connect the dots. Some will feel overwhelmed by what is about to unfold across this bright blue planet.

I knew many years ago that all this would happen, and I predicted some of it in my other writings, in radio interviews and on web sites. Some of us want to live outside the system, and want to live our own existence. Most people feel fine being cared for by the cradle-to-grave system, because they have not taken time to think independently, for even a moment on their own. If you feel uncomfortable questioning the thoughtlessness of daily life, then you know you are coming close to the truth.

Let me share with you one of the keys to opening the doors of illusion. If it makes you uncomfortable, uneasy, nervous or shaken then you have to stay with that feeling until the feeling becomes a part of you; until the uncomfortable becomes your most powerful strength. That is my secret. It is what makes me who I am. I have never been able to turn away from, or run away from uncomfortable truths and devastating realizations. I have followed them to their source and lived with them until they were a part of me.

When you can feel at ease with the force that shakes you to the core of your being, then you are ready to pass through the magi's passage of initiation and face who you really are. For you to understand what you are about to read, you must be able to *free your mind* and look for the knowledge, no matter how uneasy it makes you feel. Having the intensity to question everything you take in is a necessary element of discernment and learning.

I can show you the star gate. You are the one who has to walk through it, if you wish to make that journey. Let go of fear, belief, doubt and disbelief. Enter the world of the astrophysicist and pass through the star gate into a world of knowledge and exploration, without control or rules, without borders or boundaries, and most important, without secrets. Neither rules nor secrets are needed. All one needs is: "harm no one" and "free your mind."

We need to think anew, begin anew, to assemble the pieces of the puzzle. Everything we humans are attached to will go through the grinder, from the new age, religionists and believers of all creeds, to the atheists, agnostics, pagans, conspiracy buffs and alien chasers. The star gate is within you, close to your heart. It is not a mechanical device, nor an isolated spot on earth, nor far away on a distant star. The star gate needs no technology. You are your own power source. In my astrological world, everything is possible via knowledge. But please understand also, there is a difference between knowing the Cosmic or shamanic passage and walking through it.

I personally come from a Celtic background, from the era of Merlin and the magi consciousness. However, many other ancient seers and sages also taught the truth that the spiritual is not a *religion* but a living science. The root of the word religion means, to tie or to bind.

The Mayan prophecies are a mixture of scientific technological knowledge based on the sacred Tzolkin calendar and visionary powers of the shaman-prophets. To date, the prophecies they made have been realized with a high degree of precision. The foundation of the Mayan world is based on their sacred calendar, or their form of astrophysics, which to this day is more advanced than any astrology we have encountered. In their world, this reality exists in the *Najt* or space-time, and is an infinite spiral in which cycles and events are repeated with a certain correlation to a previous or future time, corresponding to similar events.

The Celts had a similar framework. Merlin worked with applied high astrological calculations. Did his pupils do what he taught them? The Christ, known to us as

Jesus, the Buddha and Krishna all taught their apostles, students and disciples more than will ever be known by present-day humans. But did these followers and believers really understand what they were shown by their ascended masters? Judging by what happened later, obviously they did not.

New grid-harmonics are forming around the earth. They are teaching us the harmonics of one simple puzzle: May we dream of new realities and awaken from our dreams; may the force be with us.

"Are we in the world or is the world in us?"

Dawn of A New Age

"Forget the past, remember the future!"

~ Gary Lehar, Astrologer

Believe it or not, we are actually moving forward toward a golden age, a time of mental, emotional and spiritual sanity. The struggle we face is within the mind, heart and spirit of each individual. This spiritual struggle has already begun. It is the struggle of the old hierarchical brain seeking to maintain control of the mind. The war of the mind is being acted out in secret. It is a secret spiritual conflict, an informational warfare of ideas, conditioned illusions and thoughts.

This conflict is outside the realms of Zen magic. It has nothing to do with true spiritual awareness or the true spiritual mind. No sage will take part in this war. They will navigate around it, with style, and using the elegant skills of the peaceful warrior. The ultimate outcome will be according to universal order, for spirit always guides Zen magic.

Although the *Zen of Astrology* appears not to be as precise as the science of astronomy, its foundation is firmly based on precise and accurate intuition, and the understanding of cosmic law. Zen is based on the notion of balance. Aided by inspiration, astrology works with universal languages of art and metaphor by comparing such apparently unrelated things as stars and humans. The relationship is more real than one would imagine.

The magical reality of quantum physics reveals that

stars and humans are indeed intimately inter-connected. Astronauts who have traveled out there have confirmed this and some have written and stated it in their memoirs. The ancient truth "as above – so below" encourages us to discover what we need to know through reading and understanding the eternal signs in the heavens.

The astrologer of the future integrates the reading of everyday signs into the star language via noesis or guidance, in the very same moment the sign occurs, as he compares a transit alignment of planets in their Sabian symbol degrees to the signs he is given in everyday situations. With this technique he can triangulate a message and find veracity in it.

Modern astrology is not about what must or will be; it is about what we do and what we can do. This is a vastly different understanding of the ancient art of stargazing. The divine right of free will and choice reign supreme in the art of Zen astrology, just as they do in the technology of Cosmic science.

The signs in the heavens report that Pluto (the powerful ruler of death, transformation and rebirth) entered the centaur constellation of Sagittarius (the seeker of truth) in 1995 and will leave this sign of religion, law and order, morality, spirituality, publishing, higher learning, and freedom by 2008.

Chiron (the ruler of magic) had joined Pluto in Sagittarius in 1999. Their meeting, conjunct to an arc-minute on the night of January 1, 2000, manifested the millennium signature, which signified the healing of humanity's wounded power and pride as well as the regeneration of our knowledge. This will also bring about a similar regeneration in science, technology and spiritual faith in ourselves, our Cosmic origin and in Cosmos itself.

Chiron joined Pluto in the 12th Sagittarius degree of the Sabian symbol with its keyword, *"Adjustment"* and the Sabian reading, *"A Flag That Turns Into An Eagle."* May the eagle fly and turn into a chanticleer, a rooster, saluting the dawn of this new age, for the millennium was birthed with this signature.

In 1953, Marc Edmund Jones interpreted the meaning of the 12th degree Sagittarius Sabian as, *"A symbol of balance which every individual must maintain between the established allegiances which create the substance of his conscious selfhood and the idealized ambitions by which he is self-sustained as an identity in his own right."* In other words, the individual must maintain successful self-establishment through genuine self-expression.

The above interpretation suggests that we all, as human beings of the new millennium, must present our own version of ourselves; and particularly our spiritual nature if we are to have character.

Yet our personality ought not to conflict with whatever is necessary to survive, both physically and spiritually. Actualizing our spiritual nature must not result in physical death, nor should physical survival compromise our spirituality. This will become evident when Pluto moves through Capricorn from 2008 onwards.

Spiritual understanding is revealed and refined through the tireless struggles of the individual with situations presented moment-by-moment within the space-time experience. During this dawning millennium, the individual explores new ways of seeing and understanding. The flag that becomes an eagle, in the Sabian symbol of Sagittarius 12, suggests new forms of relating to the sacred in the Aquarian age universe.

Lined up directly with Earth, Chiron (symbol of the magician, astrologer, and shaman), joining with Pluto (icon of wealth, nuclear power, and transformation), is here to teach us a new holistic worldview. Together in the sign of Sagittarius, these two planets mirrored a new spirituality of seeing and creative methods for integrating teaching and learning, psychology, astrology, sociology, quantum physics, cosmology, Zen magic and all things foreign. This means that new professors teaching at universities with an innovative curriculum designed for holistic learning will soon meet in alpine retreats when Pluto in Capricorn prepares the new structure of society by 2020.

In essence, Chiron in 2000 was ushering in a new thought paradigm, moving from Sagittarius, *I see* to Aquarius, *I know* from 2000 to 2010 and suggesting to all of us how we must enact the change. In birth charts, which I analyze with my clients, Chiron shows where we need to claim responsibility for transformation and advanced teachings.

In a few years, when Chiron moves into Pisces, this vibration will help us evaluate and restructure our treatment of each other on a universal scale. In 2010-2012 when Chiron wanders simultaneously together with Neptune from Aquarius into Pisces – in that surreal meeting zone between belief and knowledge where the two ages now meet – the planet ruling spirituality meets the planetoid ruling high magic. During this time the famous doors of perception will open for all of us. At that time Uranus will enter Aries to blaze a brand new trail.

The new discoveries will by then be fascinating, and

the quest for coherent knowledge will be self-assertive. The fragmentation of the "Tower of Babel" type knowledge will come to an end then. There is also a potential in this time for false spirituality, false gods, shadow light and misguided or deceptive attempts to create a one-world religion in this alignment and you certainly want to beware of this type of delusion while creating your own future.

I feel that all of these changes will lead to a deeper and more balanced trusting of ourselves and creating an immovable faith in our own link with the invisible world, with the unknown, the mystery surrounding us, with the universe or Cosmos, as opposed to a reliance on what religionists tell us in schools or places of worship.

The self-survivalist attitude, like desire, is a limiting reaction of the old brain. What is important in the years ahead, as we enter the wave of new energies, is to integrate our actions with the new brain and move away from fear based survival instincts. The ancient trauma is now behind us as we envision an integral, non-fractured future.

What we can do now is come more fully into relationship with powerful aspects of our experience of existence, and we can alter our perception to see and understand the value of the experience of survival, and be grateful for that which completes our experience of the grand mythological vision – the one that is the envisioned integral future of mankind, beyond trauma and beyond survival.

All things considered, the fact that mankind survived in this Cosmos is actually a miracle. But then I have always been of the opinion that miracles do happen, once we intend them to happen, which is to live direct action, free of fear, without the limiting desire for survival.

The individual human heart in tune with spirit will replace organized religions and their structured dogmas. We will seek and find others who think as we do and form communities and new societies with a common goal: Peace, compassion, and mutual understanding.

We will take steps to heal individual and group power on this planet, preparing to take our place as responsible members of this interplanetary and intergalactic society. We will live *Closer To The Stars*.

Millennium Signature

"As one lamp serves to dispel a thousand years of darkness, so one flash of wisdom destroys ten thousand years of ignorance."

~ Hui-Neng

We are more than our stars. Cosmos is immensely more than its stars. Yet, the alignments of planets and stars give us an unmistakeable clue about the psychological climate and quality of a time zone under consideration. Con sidera is Latin for, *with the stars*. The planetary and star signature of 1.1.2000 is what to me defines the 3rd millennium of our time ahead, in its odd way.

On January 1st of 2000, the moon was in Scorpio, signifying that *now* is the time to face our fears and bring light into the shadows of the collective subconscious. It was a time to exert control over our reflexes, and to eliminate or let go of outworn and outgrown emotional reactions, as well as moving beyond chaotic and undermining circumstances, relationships and situations.

I am grateful that Cosmos timed the night of the millennium with a Scorpio moon, right on my own birth moon, through which we can all look into our fears and win by so doing. Victories require challenges, and without challenges there are no victories. The psychic Scorpio moon is in esoteric terms the high flying and far seeing Eagle moon. The decades ahead bring the challenge of surfacing fears with the strength to face and overcome what we most dread. The new age thus started with the rising of the Southern Cross star system in mid Scorpio as the new millennium got under way. I have to say I really like this as it denotes considerable star wisdom to emerge over time.

While we must face the form of our fears, we can manage this type of anxiety easily. Even though fear is an emotion, based on thought, creating false religions and the misunderstood notions of time, it must transform. We have to learn to think constructively to better enjoy life here. We can change our fear-based thought processes into direct action and ultimately create magical ideas that transform our lives and our planet into some form of heaven on earth. This is the time to undertake this task.

We are and we become what we think, as philosophers of all ages have correctly suggested. Our physical bodies are merely temporal vessels for our souls: The eternal portion of our being is indeed an energy of formless thought. Whether we acknowledge it or not, we create our world and ourselves by the power of our thoughts.

We need to use the mind consciously, creatively to think magical thoughts and dream magical dreams. Magic is simple; any child can do it. But as we grow up we tend to forget the power of imagination. High or Zen magic, however, is calm, deliberate, focused awareness – the art of transformation in tune with the intelligent universe. One can ask a valid question: Is it at all possible to exist without thought?

This new millennium, the world of pure intent, is all about Zen magic – or spiritual magic – the power from within to transform and to re-align the stars of fate. One can do this, by proceeding towards action without thought, which is what I mean by working with the invisible world.

This is what Pluto and Chiron in Sagittarius were showing us when we moved into the opening days of the millennium. Now we are leaving the past and our prisons of experiences behind us as we move into the future as navigators of the unknown. As unique and independent travellers in time.

Truth is the unknown. No authority of any kind will tell us how to approach the unknown, or the truth. We have reached the point where we cannot measure the un-measurable. Science, quantum physics in particular, faced with the unknowable and eternal has crossed over into the realm of magical thought where the mind of the observer effects what is seen.

Perhaps this is how the creator planned creation, the universe, leaving free space for the creativity of the human mind to explore the creation. This is the age that tells us we are not to worship the creator of all that is and that is not, but that instead we are to create as creation itself moves on.

The Vortex Theory does, in scientific ways, propose what I have been explaining in my Zen magi manner, for a while. The ending of time is in sight. It is time to step beyond the illusion of time, to reach what Krishnamurti termed, *"Freedom From The Known."*

As Shakespeare wrote, *"We see what we believe, as certainly as we believe what we see."* Formerly belief was the non-specific force driving this world, but now the opposite applies with a magical twist. In other words, we and not some *God* are responsible for what we think as we intend it to materialize in space-time. Every intentional thought is an act of Zen magic, as is every intentional act; both of which will have an effect on us and on our actual, not imagined experiences once the thought has left the mind. This is also why we wish to be superbly cautious about what types of thoughts we let enter into our minds, the pure Zen garden of our existence.

Stepping into the unknown requires some sort of unconventional approach or an attitude of daring, courage, or grace under pressure. Yet, at the same time, it is as natural and effortless as the sun rising, water flowing, grass growing or as natural as breathing. We just do it by living each day as it comes. Yesterday was not more or less unknown than the day before, yet it is now past and completed as if it has gone the way of the Trojan War or the middle ages or the explosions of ancient star systems.

I see that the future is now. There are patterns of the past reflected in the future vortex of times ahead, and there are ways we can see this future time; as if we had come from the past into a future yet to be to rectify an outcome before it has formed. This is another form of Zen magic. We will leave the notions of time behind us as if they were mere motions of a galactic clock face, a thought dial made of planets in motions around the 360 degrees of the Zodiac.

Most people assume that we cannot shape yesterday. But using the mind we can give it a different interpretation, which means how we see ourselves effects the past, the present and the many tomorrows. The future will look just the way we think it will based on the creative power of our thoughts. How you perceive yourself is the unfolding of tomorrow. That, in essence, is high magic, which includes predictions, divination, and the art of change, also known as transformation.

What we experience as our existence is based on our

own inner and outer behavior. If we are violent we create the worst possible future for ourselves. The violent mind creates problems, wars, famine, upheaval and disasters. Then we encounter our own self-created trials and unhappiness. We need to know and understand what lies ahead in realistic terms of the extended now, through understanding our own behavior and actions.

We live that which we create. So why not imagine, think, dream, and visualize such magical outcomes as love, health, wealth, shared prosperity for all, a united peaceful clean world, a world free of conflict. Why not imagine a planet with free energy technology? If the billions of people on earth would think such thoughts, we would be well on the way to experiencing an economic revolution in a millennium golden age.

According to high tech astrologers, the last time the planets aligned as they did in January 2000 and the few years that followed, was during the reign of Thuthmosis III, the Pharaoh under which the Egyptian civilization experienced its height of glory. There may be golden ages, but there are no ordinary moments according to philosophers of the Socratic school. Every moment counts and is important. Live in the moment. Focus on what is happening now. Expect the unexpected! *"The world is ruled by letting things take their course."* Lao-Tzu.

We cannot change some events such as the track of a predicted hurricane, or so you may think, but perhaps we can indeed. We can change our thought, our attitude and our reaction to these events. We have a choice to let our fear paralyze us, or we can prepare for it, move out of its path. Nothing is cast in stone. There is always the variable of free will and the possibility for change in harmony with what some call Divine order. Are we willing to take the necessary risks? Do we have the courage to venture into new galaxies of space-time and thought while aware that most of the three-dimensional world around us will perish with the old brain?

The central problem for humanity remains the intelligent application of technology. Mankind has thus far proven to be incapable of handling hi-tech toys. Will we use thought to invent new bombs or to invent new healing methods? Is it our future to conquer other parts of the world or outer space? Are we here to heal and to teach or to destroy each other?

Are we using tools like the Internet to bring the people of this world closer together in peace, or are we using the Internet to promote deception? Are we showing our children how to assemble destructive thoughts or how to educate themselves intelligently? Are we going to develop *blank slate technology* to defend planet Earth or are we going to blow up the planet in an alien war game repeating the conflict on Mars a while back? Therein resides our challenge and our chance to change.

The challenges that face us at the dawning of this new millennium are mirrored in the position of the planetoid of magic, Chiron, seen by some as the death star. Chiron was conjunct to the transforming avatar Pluto in Sagittarius when the 9/11 psychological operation went off the ground. Chiron and Pluto started their journey jointly in Sagittarius at the same spot in 1999-2000. These two heavenly bodies were also conjunct, but in the sign of Leo, during December 1941 when the United States entered into war with Japan.

In 2010, in the constellation of Aquarius, Chiron will

conjoin Neptune, a cosmic channel of spirituality. Chiron and Neptune last conjoined in Virgo when the United States dropped an atomic bomb on Japan in 1945 with devastating results. The influence of Chiron can heal or wound; teach to transform and produce magic or instruct for demolition and effect death. In the Aquarian age of the World Wide Web, we can, for example, either disseminate ideas for spiritual growth or provide the plans for constructing nuclear bombs. Chiron is thus the planetoid to track, while the choices for right action are ours.

While retaining our trust in the universal plan and purpose, we need to develop a new faith in ourselves in tune with the guidance of the force some call Spirit. The greatest challenge we face at this time when Neptune reaches untried alignments to other planets is one of opening ourselves to new realities, and focusing the enormous and unlimited, creative power of thought within each of us. Untried alignments mean that when Neptune will be in Pisces, after 2012, it forms geometrical aspects to other planets, such as Pluto, Uranus, Saturn, and Chiron, that we have not seen as far as recorded human history goes.

I would imagine that a Zen-like enlightenment will set in globally, as if a Cosmic light switch was turned on and the darkness of thousands of years will dissipate, gradually with Neptune, very much like a dawn after the night ends with timeless wisdom becoming available to everyone. I envision this to manifest like Zen Master Dogen writing about the moon reflected on the water:

"Enlightenment is like the moon reflected on the water. The moon does not get wet, nor is the water broken. Although its light is wide and great, the moon is reflected

even in a puddle an inch wide. The whole moon and the entire sky are reflected in dewdrops on the grass, or even in one drop of water.

"Enlightenment does not divide you, just as the moon does not break the water. You cannot hinder enlightenment, just as a drop of water does not hinder the moon in the sky. The depth of the drop is the height of the moon. Each reflection, however long or short its duration, manifests the vastness of the dewdrop, and realizes the limitlessness of the moonlight in the sky."

Designer of Star Magic

"Every man and every woman is a star. Every number is infinite; there is no difference."

~ Aleister Crowley
The Book of The Law, 1904

Astrology and cosmology are necessary disciplines of being alive, disciplines we have to master and comprehend in order to better see and understand the past, present and future. Which on another level is merely the motion of space. To most people stars and planets are mere dots in the sky. To sea captains and astronauts, they are vital navigational tools and aids for finding the way. To poets, they offer heavenly inspiration. To astronomers and cosmological astrologers, like myself, stars are instruments which were created to harness *time* – to structure the un-measurable and infinite.

God invented time so nothing would appear to occur simultaneously, so that there would be order instead of chaos and so we could experience events and developments chronologically and thereby learn and grow spiritually. Planets and their effects, or non-effects, on human beings has been a discussion I have had with many scientists and religionists over many years, to the point when I turned to my inner Zen magic: *"So sorry I offended you, of course you are right."*

Most scientists do not want to understand what I know. Most religionists cannot understand it, and others choose to ignore it. The most advanced of them all who corresponded with me on these matters is Russell Moon – the discoverer and inventor of, *The Vortex Theory*. He did want and was able to understand, and he wrote back to me the following about this subject after we discussed it for over a year. Below is an excerpt from: "Proving The Vortex Theory" by Russell Moon:

Many years ago I did not believe in what has been classified as the paranormal: including astrology. However, with the discovery by the vortex theory that gravity appears to be created by regions of less dense space surrounding protons and neutrons, then Einstein's "bent space" surrounding stars and planets is really massive, spherical region of less dense space that diminishes in density according to the square of the distance from the planet or sun. This creates some intriguing possibilities.

It creates some intriguing possibilities because it reveals that not only is the space surrounding a person distorted in many different directions at once by the sun and motion of the planets [as is already believed to be happening by gravity], but also, because this less dense region of space is in motion – the speed of the rotation of the sun and planets creating these distortions and their magnetic fields or lack of magnetic fields must now be taken into consideration. According to Newtonian Physics, gravity is purely a function of mass. So unless a star or planet has an uneven distribution of matter within its geometric structure, the gravity is uniform [the center of mass coincides with its geometric center]; this means its effects upon other suns or planets is uniform. However, because the vortex theory reveals that the gravitational attraction of a star or planet is a function of all of the less dense regions of space surrounding all of its individual protons and neutrons [note: the bent outward regions of space surrounding the electrons are neutralized by the bent inward regions surrounding the protons], and since these bent inward regions of space possess individual motions [both linear and circular], there is much, much more to the "gravitation effects" of planets and the sun that originally proposed and presently believed.

In fact, there are many, many miniature currents and eddies within the individual "Gravities" of each planet. So much so that once technology is developed to study these effects more closely, it will be revealed that the "gravity" of the individual planets have a much more pronounced effect upon other matter in the universe than was previously suspected.

One of these effects will be seen in the flow of electric currents. Although the electromagnetic force is much greater than the force of gravity, when the currents and eddies within the individual gravities of planets are discovered, it will then be understood that these motions have effects upon the spin states and the motions of individual electrons in current flow. Because the "consciousness of our physical body" is created by electric currents flowing through neurons, the individual planets will effect these flows. Consequently, depending upon the position in the heavens and the distance from the earth, the effect of each planet upon the functioning and well being of our physical self will either increase or diminish.

* * *

Ultimately, when a person is born, the effects of the individual planets that are highest in the sky and hence closest to the individual [note: when a planet or the sun is behind the earth, its individual effects will be diminished by the earth's shielding effects] will become part of the individual's subconscious "environmental conditioning". And just like the conditioning of any environmental effect [such as Pavlov's bell causing the dog to salivate], the position of the planets could very well cause subliminal effects within the individual that in turn directly effects the way he acts and reacts to the everyday events he faces – causing him to feel elated or depressed, allowing him to respond successfully or unsuccessfully: and ultimately, to succeed or fail.

So you see, there is and someday it will be proven that the planets and their positions have a direct effect upon individuals and humanity as a whole. – Russell Moon.

The stars do not force us to be anything other than what we are, and yet they can guide us by displaying the secret and esoteric "writings on the wall" of the celestial firmament. Mapping the future is what this knowing is about. To me esoteric knowing has to be useful, or else forget it. *"The journey of a thousand miles must begin with a single step."* Lao Tzu

Many readers have asked me how one can read astrological charts the way I do. I have to admit I do not read a chart as such, I see the energy of the stars and planets combined in a moment of birth applied to an event or moment in transit as well as over long stretches of time. It is like seeing a clock face and knowing the whole business of the clock and how the face will look like an hour or nine and a half hours later.

How can we decipher the encrypted code that Cosmos

left us before it set events in motion? And why is it possible to make informed guesses from this celestial activity? High magic is the art of accurately reading star-planet movements. It is the art of the possible as opposed to the preconditioned and fated. Spiritual astrology as I defined it is the clock of destiny; it is a road map to fulfillment. We must pay attention and watch this charted course very carefully as we enter the third millennium with its promise of new beginnings, and with its inherent dangers.

Let us see if we can find a path through the maze of space-time and discover where we came from, where we are headed to and who we actually are. As we can see, the world did not end with the year 2000, nor was there a societal or computer meltdown as many wanted me to predict. However, the world will change and it will change in ways that defy present day logic.

The concept of *Y2K* associated with the year 2000 had to be seen as a type of informational warfare designed to keep humanity fear-based. At the same time it is a powerful spiritual metaphor for a massive upgrade into a new form of human consciousness. We need to change the collective programming in order to see further and evolve to take responsibility for our own transformation. We need to increase our psychic awareness to include an understanding of the overall picture, our possible future and the remembrance of who we really are. The age of Aquarius posits precisely that question: Who are we?

During the age of Leo, 13,000 years ago, when the architects of the Pharaohs built the beautiful majestic sphinx, the lion body with the human face, and the pyramids in Cairo – humanity asked, "Who am I?" While the age of Leo focused on the Pharaoh and his infinitely clever magician or high priest Thoth, the age of Aquarius is concerned with the whole of humanity as one group.

On a spiritual level, this symbolises the switching of the microchip in the collective memory from *me* to *us* and moving on with this new perspective. At an intellectual and lower level we can see the psychological operations of the fear mongers at work. At *Level 11*, my level, we knew what would come. Now is the time to create our new lives, as the three dimensional world spins out of control. Infinity and universality can be looked at and applied via understanding the alignments of the planets.

The years ahead see quite a number of very rare alignments. They are teachings about how we can transform our world into a magical life. Let us move quickly one more step back before we jump forward. We must understand how we measure time in terms of outer planets conjunctions that denote beginnings and endings of cycles. Conjunctions are moments when two or more planets meet in one degree of the Zodiac.

Neptune–Pluto conjunctions

A true measure of time that has taken place only five times in two thousand years. It occurred in the year 82 BCE in Taurus (reign of emperor Titus transferred to his brother Domitian while Mount Vesuvius erupted, burying Pompeii; Roman conquest of British island); in July 411 AD in Taurus (the Goths, led by Alaric, sacked Rome); in May 905 AD in Taurus again (reign of Leo VI, the wise, Byzantine emperor); in April 1399 in Gemini (crusade of Nicopolis, Ottoman empire halted by invading Mongols); and the last one was in April 1892, again in Gemini.

The next will be in about 400 years. Neptune conjunct Pluto symbolizes a shift in consciousness, when spirituality (Neptune) merges with transformation (Pluto) to transcend into new paradigms. The first two conjunctions coincide with the beginning and the end of the Roman Empire, while the second couple of conjunctions mark the medieval and the renaissance awareness cycles. The last conjunction coincided with the term of Benjamin Harrison, the resignation of Otto von Bismarck, and the Sino-Japanese war, where China was almost annihilated by Japan. It ushered in our *Age of Aquarius* consciousness, and marked the industrial revolution.

Uranus–Neptune conjunctions

Rare 171 year cycles. Recently: November 1307 in Scorpio, when the Templar Knight's commanders were arrested; December 1478 in Scorpio again, rule of Lorenzo de Medici, the Magnificent, when art and science reached beyond religion; in October 1650 in mid Sagittarius, Oliver Cromwell's rise to power, reign of the Sun King, Louis the 14th, (the mid Sagittarius degrees see Pluto currently transiting, while this book is being created); then again in December 1821 in the first degree of Capricorn, at the beginning of the reign of the 6th emperor of the Manchu dynasty in China (Tao Kuang); and the last one was in October 1993, at 18 degrees Capricorn, Yeltsin suspends the Russian parliament. The next is due around 2170.

Uranus–Pluto conjunctions

I select only a few from a long list: March 1090 in the first degree of Aries; March 1344 Aries; (Hundred year's war between England and France); October 1455 in Leo (fall of Constantinople, War of the Roses in England); January 1598 Aries (edict of Nantes granting French Protestants freedom); September 1710 Leo (reign of Peter the Great in Russia, England and Scotland joined under the name of United Kingdom); March 1851 Aries (David Livingston's African expeditions begin); 1965/66 Virgo: Vietnam war. The next will occur about 2100 in Taurus, upheaval of earth resources. Uranus-Pluto conjunctions mark the beginning or ending of social movements, whereas Uranus-Pluto oppositions delineate the precise moment of social unrest and faith revolutions, as we will witness in the years 2046-47. We will analyze this later.

Saturn–Uranus conjunctions

Cycles of about 47 years. The last few occurred in January 1762 in Aries, reign of Catherine the Great in Russia after her husband's death; in November 1805 Libra, Napoleon defeats Russia, Austria, Prussia and occupies Berlin; March 1852 Taurus, Napoleon III reign begins; September 1897 Scorpio, Greece and Turkey at war, annexation of Hawaiian Republic by the USA; May 1942 Taurus, Battle of the Coral Sea; and October 1988 at 28 degrees Sagittarius (Galactic Center), Chile rejects Pinochet.

The next are due June 28, 2032 at 28 degrees Gemini, across from the Galactic Center; and October 21, 2079 at 28 degrees Capricorn. What headlines will make news? Saturn (old) conjunct Uranus (new) symbolizes the imbalance between ruler and people, democracy, rebellion, restriction, status quo vs. innovation. Humanity breaks through society's structures by keeping what is healthy and disregarding what no longer works.

Opposition cycles are interesting to watch in this respect, as the following list shows: The Bolshevik revolution (November 1917) occurred during an opposition between Leo-Saturn and Aquarius-Uranus; Scorpio Sun and Leo Moon on Saturn acted as T-square triggers. The next Saturn-Uranus opposition (Virgo-Pisces) is from November 2008 to September 2009. In July 2010, Saturn in Libra still opposes Uranus in Aries. The event takes two years to perfect and can manifest in many ways due to the different signs.

The July 1789 French revolution began during a Jupiter-Uranus conjunction in Leo. The king was executed the same day as the Uranus (Leo) – Pluto (Aquarius) opposition, on January 21, 1793. The next Uranus-Pluto opposition was in Sagittarius-Gemini from 1900-1902: Boxer rebellion in China, Philippines revolt against USA, the pot boils over in Moscow, murders of President McKinley and King Alexander I of Serbia.

Saturn–Neptune conjunctions

36 year cycles, but seen in the same signs they follow in a 321-year cycle. I selected only a few pertinent moments in mankind's history: September 1307 – in a stellium with an Uranus conjunction – in Scorpio, secret arrest of all the Templar Knights; in May 1523 Pisces, publication of The Prince by Machiavelli, introducing real-politics; August 1630 Scorpio, Galileo deals with the Inquisition; August 1917 Leo, Lawrence of Arabia – a Leo – takes the port of Aqaba from the Turks in his brilliant Bedouin coup; July 1953 Libra, Elisabeth II coronation, Egypt declared a republic; November 9, 1989 Capricorn,

Saturn's rulership sign, to the day coinciding with, and so symbolic of, the fall of the wall in Berlin. Note also that a Venus-Uranus conjunction formed within a few degrees the same day: super Capricorn stellium, as the wall fell.

Saturn: time, duty, organization, hierarchy, and structure. Neptune: spirituality and illusion, ruling visions. Saturn – Uranus – Neptune stellium conjunctions are beyond rare: Before 1989/90 in Capricorn they were joint in 1307 in Scorpio (Templar's mass arrests by king Philip le Bel), and in 1777 BCE in Capricorn, (first Babylonian dynasty, marking the end of the Middle Kingdom in Egypt). The next Saturn-Neptune conjunction occurs on February 20, 2026 in the first degree of Aries, set off like a stick of dynamite and triggered on the same day by a hot Aries moon.

In June 2025, Jupiter in Cancer will square the almost conjunct Saturn-Neptune within 5 days and less than one degree: important events will happen. The last Saturn-Neptune conjunctions in Aries were in March 1703 during Queen Anne's War; and, further back in March 1380, marking the English Peasant's revolt and the start of the reign of the insane French king Charles VI. Next in line are: June 7, 2061 at 20 degrees Gemini; then August 27, 2096 at eight degrees Virgo.

Saturn–Pluto conjunctions

An amazing list of events: 1212 Leo (Children's Crusade); 1248 Scorpio (7th crusade); 1284 Capricorn (reign of Philip IV, the Fair in Paris, introducing the first absolute monarch in Europe; 1318 Pisces; 1350 Aries (the Black Death decimates Europe by half of its population); 1382 Taurus; 1414 Gemini (battle of Agincourt); 1445

Cancer; 1480 Libra; 1518 Capricorn (Protestant reformation under Martin Luther; first circumnavigation of the globe by Magellan); 1552 cusp Aquarius/Pisces (The religion of the region shall be that of the ruler); 1595 Aries; 1616 Taurus (foundation of the Romanov dynasty); 1648 Gemini (peace of Westphalia, with Calvinism chosen as German state religion); 1680 Cancer; 1713 Virgo (end of Spanish succession war); 1750 Sagittarius (reign of Francis I, as Holy Roman emperor); 1786 Aquarius (constitutional convention in Philadelphia); 1819 Pisces (Bolivia declared a republic); 1851 Aries; 1883 Taurus; 1915 Cancer; 1947 Leo; November 1982 Libra (rise of KGB head Yuri Andropov).

The next will be in early 2020 in Capricorn rulership; 2053 in Pisces; and 2086 in Aries. Saturn-Pluto conjunctions stand for structure dissolution and society change. In Capricorn it heralds total upheaval, from the very top on down. Nothing will ever remain the same then.

Jupiter–Saturn conjunctions

A most mysterious phenomenon has been noted: All Jupiter-Saturn conjunctions from 1842-1961 took place in earth signs, and all 7 Presidents elected in years ending with zero died in office as Jupiter-Saturn were in earth signs: Harrison, Lincoln, Garfield, McKinley, Harding, Roosevelt and JFK. Reagan was elected in 1980, during a Jupiter-Saturn conjunction in Libra, air sign; and got shot in his lung. Jupiter-Saturn was conjunct in earth sign Taurus in May 2000.

On November 7, 2000, Saturn (father) was on fixed star Algol (Pile of dead corpses) in Taurus. We may guess how Bush fares with this, as the ballot recount (which I predicted in writing in December 1999) casts a shadow on his leadership. As in 1960 (JFK), 2020 will see Jupiter-Saturn conjunct in earth sign Capricorn, election year.

Pluto will be in the Capricorn cluster during all of 2020. The exact Jupiter-Saturn conjunction perfects only on December 21, 2020 (after the election), at zero degree 29 minutes of Aquarius. Nostradamus predicted an assassination for Easter 2021: Will the cycle (some say it is an Indian shaman's curse) continue or be broken? Do I care? No.

Three special and spectacular super stellium clusters of greatest rarity in the near future are: April 29-30, 2011 – Six Planets in Aries: Moon, Mercury, Venus, Mars, Jupiter, and Uranus, all squared by Pluto in Capricorn, while all opposed by Saturn (Ruler of Capricorn) in Libra. To me, that moment in time is symbolic of a very fast and hot new beginning, with maximum unrest. It could be the time Nostradamus calculated as the beginning of an eleven-year war zone.

December 26-31, 2019 – Six Planets in Capricorn: Sun, Moon, Mercury, Jupiter, Saturn, and Pluto. The last days of 2019 will open a cycle of three Capricorn loaded months, with Saturn in rulership of its horned goat sign, culminating in mid March 2020 in a conjunction with Jupiter and Pluto. I see this as the apex, or zenith, of an old society structure. Analyze the Saturn-Pluto pow-wow, and check in the above pages what type of event transpires historically in such alignments.

June 7-9, 2032 – Six Planets in Gemini: in an extremely rare, almost conjunct, stellium alignment: Moon, Sun, Venus, Mars, Saturn, and Uranus. This date will coincide with the push for freedom of expression.

Pluto in mid Aquarius trines the cluster, as Neptune in Aries sits in the midpoint of the alignment made between the Gemini cluster and Pluto. The last time Pluto was in Aquarius, was during the French, British and American revolutions. Pluto in Aquarius tends to open up structures globally. Neptune in Aries dissolves ego-based belief systems.

Pluto–Cosmic Outcast or Grail–Bearer?

Pluto was in Aquarius in the 5th and 2nd century BCE; then in the 1st, 4th, 6th, and early 9th, in the late 13th to early 14th, in the 16th, and the late 18th centuries, and will be there again from 2023-2043. Pluto in Aquarius signifies a shift and transformation of consciousness for an upgrade of mankind's awareness and places man squarely in the larger questions of humanity. The water bearer, we remember, deals with the issues of a global and humanitarian vision, faith, groups, associations and causes, whereas Pluto is the planet bringing about transformation via depth analysis, death, birth and rebirth. It rules over the collective subconscious, wealth, and nuclear power. In Aquarius, Pluto transforms views of social, global awareness in radical ways, as the historic list of events in the past shows clearly.

Major alignments which Pluto will make to outer planets, while in Aquarius, include a psychic sextile (60 degree angle) to a Saturn-Neptune conjunction in early Aries, in spring of 2026, a beneficial trine (120 degree angle) to Uranus in early Gemini, in summer of 2026, boding well for a good start in matters of upgrading consciousness globally. June 2028 to January 2029 sees a Saturn-Pluto square, questioning resources, as Saturn is in Taurus. Saturn in Leo opposes Pluto in October 2035, questioning authority, followed by an added and fixed T-square from Jupiter in Taurus in February 2036. A world meeting, as predicted by Nostradamus, shall take place during that time. From September 2046 to June 2048 (with August 2047 in between), Uranus in Virgo opposes Pluto in Pisces, for years, when I believe we will see the ultimate turnaround of humanity, or its demise.

Jupiter–Pluto conjunctions in Aquarius:

All or nothing, 240 year cycles of mystic power. To end my lesson in history of time and planetary alignment comparison on a note of irony, let us look at the last Jupiter-Pluto conjunctions in Aquarius: 799 AD, proceeding the moment by one year when Charlemagne was crowned emperor of the West by Pope Leo III. 1061: A college of cardinals must elect Popes. 1298: Philip IV takes the Pope hostage. 1534: Henry VIII declares the Act of Supremacy and himself above the Pope, while Jesuits form their own society under Loyola. 1796: Napoleon's rise to power.

The most spectacular, memorable and ancient Jupiter-Pluto conjunction in Aquarius occurred 64 BCE, when Emperor Nero annihilated Rome by setting it on fire. The next Aquarian Jupiter-Pluto conjunction will be on February 4, 2033, to herald a new time, manifesting a benevolent grail-bearer, or an open society with a council of elders elected by the Aquarian people of the world. Nostradamus predicted for 2033 a chemical war, rising prices, and two revolutions followed by years of starvation. End of story.

No matter what the planetary alignments suggest,

ultimately the outcome of what they might predict rests upon the choices we make, individually and collectively. The outcome rests on how to use the energies the planets channel onto the earth plane. Consequently, our choices create the future that will become mankind's history. Let us remember once more at this point that if we can predict the future, we can change its outcome by choosing a new future in the present. Since potential futures lie in a state of sleep to be awakened in the present moment we are the architects of our own design.

As this book appears, we see Pluto in Sagittarius conjunct Jupiter in rulership, at the Galactic Center. In 2032 – Pluto in the middle of Aquarius – the inner planets and the sun will line up with a rare Uranus-Saturn conjunction in Gemini, across from the Galactic Center, in the sign of the wind that moves around and beyond the limitations of thought. The future and untried (also known as the unknown) meets in a liberating way with the form of the tested.

This major conjunction speaks to us of a breakthrough moving to a new understanding. Freedom from the known is active at this point. The new mind will be accepted. The value of one life then is equal to – if not more than – the lives it touches. The happiness of one is in independent origination to the happiness of many.

The key Sabian degree is Gemini 27: *"A Gypsy Emerging From The Forest Wherein Her Tribe Is Encamped."* It seems as if the High Elves will make themselves known to those who can see through the dimensions. Resources and enthusiasm will be available with many options open for the people alive on this earth at that time. New languages, new forms and new colors will have emerged. Expanded mental horizons are the result of us knowing our own human potential. Realizing this potential is what the age of Aquarius will propel us into.

When does the age of Aquarius – also known as the age of the group knowledge – begin? First we look at a scientific explanation, in Aquarian manner, to see what an age cusp is defined by:

As the vernal equinox (Sun into the first degree of Aries) enters a new constellation, a new age starts. This happens about every 2,160 human years. Now imagine this in Cosmic reality and you see why this question has the sages mesmerized. Also know that in reality we have more than 12 signs, at least 13.

The precession of the equinoxes – a movement backward through the zodiac – produces a shift from Pisces into Aquarius that covers the degree zone Pisces 1 to Aquarius 30. This may sound like nothing but it takes its time. Due to the less than clear cut boundaries of the constellations, there is a vast spectrum of dates given as to the beginning of the new age, from as early as the 18th Century to as late as the 22nd Century.

My own seer interpretation is just as scientific. Psychologically speaking, when we move, as one humanity, from blind faith and the belief systems of the past two millennia to the human knowledge and trans-human reason of the future, we could say that we have truly entered the vibrational levels of the age of Aquarius. In this way it is clear that as you hold the first 2006 edition of this book in your hands, we are still putting the finishing touches to the age of Pisces.

The location where we are looking for the shift into a

New Age to occur, is somewhere between Cetus and Pegasus, where ecliptic and celestial equator meet in the vernal equinox. We are probably going through one more last degree, or about one human lifetime to reach this point in time space. Keep in mind that the precession of the equinoxes is formulated as 72 human years per one degree, which is a correct measure of time space.

An inspired and insightful study of extra-terrestrial time capsules overlaying human history into planetary and star alignments combined with an analysis of ancient myths in Egyptian temples, texts and other sources – taken from the invisible world as well as from Tibetan knowing – shows that the masters of time knew and measured the ages back to the old kingdom.

There are some hints that this knowledge went back to the age of Leo, and it can be proven that in the age of Gemini which ended somewhere near 6,140 BC we have a valid time marker. In any event all this was long after Atlantis was gone.

A finer calculation points to the fact that the Egyptian magicians might have considered that the concept of the great return begins at the age of Leo when the stars in Orion were at the lowest declination. The Great Pyramid and the Pharaoh's chamber were used to send the Pharaoh's spirit towards the Orion constellation. Which was used as a space port, I would add.

You will find the constellation Orion, the hunter, near Gemini now. By use of modern astronomy software one can estimate the age of Aquarius to begin approximately around 2,070 AD when a half precession cycle of 12,920 human years shows the Orion constellation at its highest declination and its maximum altitude at the Meridian.

These considerations go beyond the scope of this small book. The *Star Zen* magical answer is simply this: The age of knowledge dawns in the circuitry of the human DNA body as you read this book. Considering also theories by reputable astronomers, I would situate the beginning of the new age in astronomical terms in the zone between 2045 and 2090, but probably closer to when Pluto is in tropical Pisces and therefore in sidereal Aquarius, while opposite Uranus, for reasons of aesthetics.

Uranus will enter tropical Aquarius again in 2080, but this does not really have anything to do with when the *Age of Aquarius* begins, although it sounds nice once every 84 years. A question of the ages is a consideration of geometry when we care to contemplate this matter deeply, using the Zen of seeing. Why do I say this?

The Zodiac of Dendera in the Temple of Hathor (from around 100 BC) was a motion of time clock measurement that marked (presumably correctly) the precession of the equinoxes from the age of Pisces back to the age of Taurus which started about in 4400 BC, give or take twenty human years, or one generation. One could – based on this valid approximation – estimate the age of Pisces to have started near 60 BC. Thus the age of Aquarius starts at about 2100 AD, as the 22nd Century of our measurement begins.

Literature with compelling arguments sets the date of 221 AD as the key moment in the galactic clock of destiny when both the sidereal and the tropical zodiacs were in alignment, or congruent. From this point one calculates the first star in Pisces to cross the vernal point at 111 BC, which would place the beginning of Aquarius at around 2,060 AD, when Pluto is in that region in Cosmic space.

Through my inner self alignment I have settled this matter by using an alignment of planets last seen in 5 BC when most sages agree the trans-human Christ was born. This alignment is repeated and active again in June/July 2048, and it is highlighted by a set of two Sabian symbols lining up over the Pluto-Uranus opposition; an opposition that in its own right symbolizes the shift of the ages, and that in and by itself marks to me a transition point of relevant proportions: Pluto at Pisces 9: *"A Jockey Intent On Outdistancing His Rivals Becomes A God of Speed."* Uranus at Virgo 9: *"An Expressionist Painter At Work."*

"When you pass through, no one can pin you down, no one can call you back." ~ Ying-An

Activate Fluid Intelligence

Star Knowledge Guards Itself

Evolutionary Earth Changes

Closer To The Stars

Trends, Forecasts & Nostradamus

Transitions of Beings

Extra-Terrestrial Guidance

Celtic Cetacean Delphinus

The Special Ones

Star Knowledge Guards Itself

"When we ponder and wonder why
Always we turn our eyes to the sky.
For skies are the eyes of powers unseen:
Keepers of wisdom inside of this dream."

~ Michael Donovan,
Letters Upon The Mast

Zen of Stars... How? There are different kinds of astrologies. The accurate and new astrology – the science of Cosmos is in fact multi-dimensional and will be introduced to humanity at the right time. This subject is so complex it cannot be summarized in a single book. Suffice it to say that planet Earth, and other planets, have at times gone through what we call catastrophes or upheavals.

There is no point knowing in detail how, why, or even when. It happened and it will happen again. For some reason, and I do not know the reason, the key knowledge that survived throughout all of time, all ages and all spaces of Cosmos, was about how to *navigate*.

To navigate from anywhere to anywhere you need not know anything about time. Time as we know it here does not exist on that level. What exists is *motion* and s*pace*, in harmony with the forces that move the whole thing about and around. Creation is self-organizing, and in many ways un-predictable.

Cosmos is as chaotic and creative as human beings are. Creation is ultimately and most probably akin to its

91

human child. Regardless of what scientists believe to be true, creation and Cosmos are so complex, so chaotic and so alive it defies any mathematical formula. I doubt creation would tolerate to be ruled simply through mathematics.

For me, the stars are the guiding map of creation. They serve as *Icons of Destiny*. Unfortunately, what we know as astrology today is only a fragment of what was known by the ancients. The knowing of the real star constellations has been morphed over the ages and their true meaning hidden. The *Zen of Stars* is still concealed, as it protects itself. I have seen and studied cosmologies that talk of as many as three dozen star signs, but thirteen or fourteen at the very least, if we take Ophiuchus and Arachne as signs. The science of the ages is still a mystery.

Considering the vast knowledge hidden from us at this point it may seem that we are lost in space. The good news is we are not lost. We know exactly where we are at in the larger scheme of things. A *silent navigator* is built into our DNA and the *inner navigator* is the deathless spirit-being that inhabits Cosmos.

One part of the solution is, we will discover or create a new language defining where we came from and where we are heading. One could imagine that this language is actually very old, timeless. The new language will appear as a new form of sacred geometry or divine-mathematics.

Somewhere inside star constellations resides a form of consciousness, some sort of navigational tool that forms and fuels its own knowledge, which to me is the Zen of knowing seeing its own future. This future is the place we are headed to. The formula is: Human being (X) plus Cosmos (Y) creates F (Future). We are living co-creators, albeit asleep and unaware, at this present time.

Dr. Edgar Mitchell told me that we will not dis-cover the future – we will create it. I have the feeling we can only create that which is inside of us... As Cosmic beings and star children we are not to worship God so much as we are to be the complete manifestation of our creation. We are supposed to manifest that for which God created us namely, *to create*...

In order to help us on our way to achieve this aim the creator created stars and planets into which is encoded the Cosmology or the history of the origin of all that is. The word for Cosmos or space in German is: *Das All*, literally meaning, *The All*. Once we figure out that we are the origin, we are on our way to becoming what we were meant to be: All one, Cosmos, Cosmic mind being. That is what we are. We are not separated from it. This is what the coming transformation and changes are about. It is the challenge ahead, or the *passage to peace*.

The seed is its own protection. The clans of light operate with an energy emanating from the heart of Cosmos, and are able to step beyond the elite hierarchy with a technology of peace known as orgone energy, emanating from the heart of every enlightened human being.

Non-linear time... The high elves, the wise and gentle forces of the invisible world, point to nature as the origin of language. The trees of the elves are our teachers. They taught me how to remain part of the invisible world and walk in this world. They say it will take a critical mass of people to show symptoms of knowing, before we who work with the guidance can manifest a clearer future. For now, we work from the back ground, moving forces into place.

Non-linear time is beginning, and this is why we have this anticipatory guidance and memory occurring in this Jupiter–Saturn–Mars–Neptune alignment, the *Grand Cross*. Now we are witnessing it from inside our heart, as the moon in Sagittarius is about to meet Mercury and Pluto close to the Galactic Center. When you look at the famous scene in 2001: A Space Odyssey, by Stanley Kubrick, where astronaut Bowman (the archer) is initiated beyond the infinite, you get a feeling about what is inside of this Galactic Center he aims into.

Today I saw a different world in my inner eye of dreams, a world with peaceful capable survivors. In the coming years, leading up to and into 2007-2009, it would be wise for all of us to begin to protect the inner circuitry, as we also mirror Cosmos and the earth in our own way. Some people will be awakening now, while others will be falling dead and not just from some pandemic bird-flu. Many will pass on because they are not prepared for adjusting to the frequency of the necessary inner changes.

Plants offer a spiritual as well as physical protection. We can use plants and crystals to create orgone generators, similar to the creations of Wilhelm Reich. Those creations can be used in homes and offices and we can tell them (by charging them) how to regulate the stream of things.

I sense that this crack in the wall, Pluto in 2008 in Capricorn 1, is going to be something much bigger than a physical tsunami or a volcanic eruption. The physical Earth is resonating to a psychic mind alteration in the consciousness of all planets. This is new resonance is being felt by people now. It takes courage to stand outside of the shock and trauma, and hold steady to a greater vision of

Cosmos as the world enters into chaos. Today most of the people of the world are asleep. They have forgotten who they are, as greater plan, as yet unseen, remains hidden in the growing chaos. This change is unfolding in the space beyond what we see and know as our everyday lives.

Many thousands of years ago, peaceful cultures on earth, developing their own skills in their own original ways, created light energy. They birthed what has become the Northern European culture. Much later more war like lower races developed an extreme arrogance, spreading their doctrine though murder and fear. That is the simplified version of the theosophist and Tibetan, *Secret Doctrine.*

Most individuals, especially when they gather together in groups, love to worship and develop the ego or group self. But their thoughts are not balanced. In fact they are extremely unbalanced. The violence destroys the collective psyche of the people it inhabits. Because they are so out of sync with creation the problems only get worse, and the violence increases. That is in essence the state of planet Earth in crisis.

The genetic manipulation of human DNA by lower-earth Reptilians, and later by other aliens, has created an even greater imbalance. That imbalance is based on conflict and violence. I mentioned several times how Jiddu Krishnamurti rightfully pointed out decades ago that belief systems are based on fear, and association with those group systems inherently breeds violence by promoting a divisive attitude.

If you want to divide and colonize a species, violence is the way to do it – violence taking apart the species from inside. To counter or combat this strategy, humans must overcome the impulses of serving only the petty self, the fragmented ego. This is the problem humans face today. Living isolated in the circle of the self creates a narrow field of vision. The isolated prison of the ego is created from the energy of the person's own mind and therefore it becomes a self-regulating inner control system. The walls of the mind prison are created and maintained by our own thought patterns.

It is essential to free oneself from this restriction. Deep down, every human being knows this is the wrong way to live. They are consciously ignoring their inner guidance as the ego grows. Each person has a powerful *inner navigator*. This higher level awareness signals to the elementary being the right course of action. To ignore the signal requires anger or violence. Thus, it is anger of the self – the conflict of *I want* opposing one's own conscience that creates inner conflict.

The ego turns away from the inner guidance system and does its own thing 24 hours a day seven days a week, living for its own desires and needs. Each wrong move is accompanied by the awareness that it is wrong. One always has the point of inner radical change close at hand. Multiply by about 6 billion times a few wrong moves, and you see mega-tons of depleted orgone energy (DOR) are created.

Psychological (psychic) pollution is equally, if not more harmful to this world than physical pollution. The negativity, fears, sorrows, conflicts and angers we create and power out every day with our minds is similar to radioactive waste. The repeated "psych-dramas" even take form, and can exist as fragmented thought-created entities, if the feelings are extreme and excessive.

In some ways we are being haunted by the thought ghosts and fragmented emotions of our past lives. Therefore it is only right that we are the ones to end this trauma and heal the past, freeing us for the now meeting the future. Those of us, who turn away from the inner navigator warnings, choosing to live for desires of the self rather than for the right action, are creating thought-monsters. This is what ego is: *Energy Goes Out* – EGO.

This is extremely suicidal behavior, as humans globally march to their appointed and final self-created downfall. The net result is global self destruction, mass suicide in a human created thought-demon cloud of mass fear and confusion.

However, if we decide to live in a serendipitous powerful way, i.e., freed from the self or ego, then we gather energy into the radiant body. This energy is ultimately self-empowering. The radiant body is able to travel into the circles beyond the isolated island of self and communicate with those other (higher) realms. We can be our own masters, but first we must move beyond the petty ego, its wants, its demands and desires.

If we live only for the self, our energy becomes depleted and at some point we destroy our own integrity. The integrity of the energetic being is essential, if we are to move into the circles of awareness beyond isolated self.

Our energetic integrity is also the vehicle by which our souls travel from lifetime to lifetime. This means those who live only for the petty self also die for the petty self. The reason to change is not in order to obtain more power, material goods or fame. *The reason to change is for love, and in order to feel compassion.* Only then does the inner navigator know the being has truly learned its lessons and is free of internal corruption.

Living wholly for the desires of ego can be compared to a computer hard drive that self-destructs via an internal self-created virus. If we further consider that other higher vibrational levels are not to receive corruption viruses, then it would make sense that there is a built-in filter process.

Who maintains this integral filter? The soul maintains the integral shield. The Knights Templar are most associated with the Holy Grail. What is the sacred chalice other than the integral soul? Integral, the root of which means *to touch*, is the holographic mind, the holographic universe.

If you take a holographic image and fragment it into a thousands parts, each fragment will contain an image of the whole picture. If you take a soul-awareness and fragment it into a thousands parts, each living fragment will contain an image of the soul. Each soul is its own soul guidance system, the inner navigator, connected to what the *WingMakers* call the Wholeness Navigator of the Cosmos. The soul oversees its own incarnation via a feed-back loop to *Cosmic Mind*.

If there is a fundamental change to a higher vibrational light within the incarnating spirit, then the soul-navigator responds and the necessary restrictions fall away. Each one of us chooses our own levels of restriction or zones of empowerment. The God-created human soul disk drive is limitless; only thought sets its own limits. If people would feel compassion and realize the inherent true power they possess, everything would change overnight. To understand this merely requires human imagination. It would only take ten percent of the population to radically change the world, through a shift in consciousness and for us to experience a major transformation.

Spanning across this present age a dark force has sought to influences the planet and humanity from the shadows. This force feeds from the fear, suffering, conflict and frustration that accompany desires of the ego. Peaceful and happily enlightened human beings will eliminate this force, giving the elite a bad headache. However, the fear embedded in the psyche of large numbers of humans is causing increasing violence and disruption. For those who are beyond fear the focus is on the oneness of mind and not on the demonic force.

It is true, unfortunately, that most of the world's population prefers daily chaos and restriction at the most personal levels. However, the personal drama is about to be cast into the earth changes drama, whether you are ready or not. The wars are created to keep the world in a state of fear and to keep planetary life unstable. The planet Earth has now reached a critical mass, fed up with playing home to six billion fear generators.

The dark psychic shadow-force inhabits spaces where there are concentrations of DOR, or depleted orgone radiance, and yet small groups and individuals around the planet are beginning to create islands of harmonic-light and cleansing. Through our actions we have the power to change what is. As the earth-transformations take effect these areas of light will come together and soon be large enough to fill the whole world. We humans, in the right state of spin, are the dead-orgone blasters, or the psychic guardians of the universe. With Jupiter in Virgo until September 2004, a massive cleansing job has begun.

If we look at world events in this way, it is easy to understand why violence is erupting all over the planet at this time. Those aligned with this dark force artificially create terror in order to try to destroy the light and the peace. The dark parasitic force feeds off negativity and fear. Positive light energy from a large enough group of enlightened individuals creates safe areas that will survive the earth changes.

With the emergence of light, the dark force will have to leave, and humans will have a chance to reconsider who and what they are. Each of us can contribute to the building of the light by asking who we are, and releasing ourselves from mind-controlling beliefs. If you look at the terrible state the world is in, you start to question how intelligent human beings can manage to do everything wrong. Chemicals, pollution, synthetic drugs and medications damage the light-body.

The increasing production and demand of material goods pollutes the water supply as well as the air. If this continues, scientists claim the earth will be a dead and lifeless planet by the year 2050. Chemical fertilizers are destroying the living microorganisms that create a healthy soil. Nuclear power stations will soon become unstable and turn into lethal and deadly hazards as they age, and as the earthquakes increase there is also the danger of nuclear explosion.

The new weapons they are creating could destroy vast areas of the earth and its atmosphere. Areas will become unlivable for many generations. If we don't act now, we will soon find ourselves living in an uninhabitable, barren dead zone. Apparently humanity is not intelligent enough to develop technology beyond its dependence on oil, and it is not sufficiently resourceful to create a type of power that is clean rather than one that destroys. We know the energy alternatives are out there, but the minds of the people are not focused on these issues. Popular mass media and computer entertainment corporations are creating products and services that not only damage the brain, but also the psyche. Humans are no longer turning to nature for guidance, and the earth is running out of time.

It would seem that all this adds up to an overbalance of negatives. However, I would ask you to consider that this may also be a purposeful concerted effort to wipe out the human populations. It seems that greed and self-interest rule the world as never before, placing us collectively in prime spot for a major catastrophe, or a series of catastrophes that will destroy our societies.

How can it be, in a world of technologically inventive humans, that we prefer to self-destruct in such obvious ways, and that our technology is becoming more polluting and destructive rather than less? Why are the mass populations fulfilling this death wish? The biggest, most luxurious and most expensive cars now make a disturbing drone noise.

The new technology we are using is not quiet, nor is it any more efficient. This means we are not really advancing technologically... and yet we are surely creating a larger hole to pay for something that ultimately destroys us. It is time that we expand our awareness to include the greater scheme of things, and peacefully take control of our individual and collective destinies.

Evolutionary Earth Changes

"Nature is trying very hard to make us succeed, but nature does not depend on us. We are not the only experiment."

~ R. Buckminister Fuller

We have entered the years of earth changes already in 2004 to 2012 – the last 8 year solar cycle of the 26 millennia spanning Mayan long count. For some strange reason it has become popular to consider that 2012 is the end of the world. It is not. Similarly strange is a notion that humanity is in command of the approaching earth changes. They are not. They are totally out of any human beings control.

Also it is thought that we are in control of the evolutionary star gates on earth, and we know what we are doing, while in fact we are in the process of destroying this cute pale blue dot called planet Earth. It makes no sense that a being said to be *in command* would destroy themselves and their own natural habitat; destroy the only home they have, including the planets food source.

We will soon experience parallel universes and realities beyond the human brain's linear *scientific* comprehension. Merging realities that will explain our own behavior and show us who is really doing the damage.

I will share with you a Cosmic history of humankind as I see it, and according to the way my other world friends taught me. Some of you may be upset, just as I was, at having wasted time at human schools studying what humankind teaches humankind, because it is all utter rubbish.

When the Roman Empire and their minions could no longer hold together their corrupt and therefore decaying but enforced colonization of much of the world, the structure collapsed and regrouped in a more insidious form that diversified into a world religion. This was the underground-alien attempt of the *Draco-reptilian* takeover. To understand what I am talking about you would have to understand the secret-cult origins of Rome and the origins of the Roman Empire.

From The Emerald Tablets of Thoth, by Doreal, it is written: *"In the form of man they [are] amongst us, but only to sight were they as are men. Serpent-headed when the glamour was lifted but appearing to man as men among men. Crept they into the Councils, taking forms that were like unto men. Slaying by their arts the chiefs of the kingdoms, taking their form and ruling o'er man. Only by magic could they be discovered. Only by sound could their faces be seen. Sought they from the Kingdom of shadows to destroy man and rule in his place."*

In this present century the whole corrupt and therefore decaying world colonization is collapsing. We are in the terminal meltdown stage until rebirth can occur, rebirth of our *human* values as opposed to *inhuman* values. This world-wide decay is not a result of the highest evolutionary process. This is not what the creator brought us to earth for. As the universe watches us with indifference; we must accept that the *creator* gave us free will and choice, to decide for ourselves which direction we are going to take.

The result of lowering human existence to the lowest

vibratory level is that our world falls apart and we become an endangered species. Human beings must be in command of their own evolution through exploring the balance of the Cosmic octaves within their own psyche, and thus building their own *psychic defense*. Humankind must understand and master the sound of their emotions.

What you feel has its own *vibratory sound*. When you become aware of this, rather than react to the feelings, then the source of the vibratory sound reveals itself to you. Feelings have to be felt-heard-seen. When you can hear the vibration as well as feel it, this inner awareness is its own meditation. This will be explored through the collective understanding of the *Akashic Records* in the Hall of Records, to be discovered in Egypt, as I predicted.

We will begin to understand our origins by studying Cosmic astrophysics, ancient scientific astrology, or what I call the ultimate science or absolute truth. When a society enslaves itself to the lowest octave, as this 4th world has done by now, it is not a sign of evolutionary development or wisdom; rather it indicates devolution and imminent death. This is part of the reason why the earth will now start to shake off its little parasites, whoever those parasites might be. To accept or to let it happen is to be suicidal, and to deny that it is happening is to be insane at best or criminal at worst.

Humanity is infinitely older than is taught. Through the manoeuvres of colonization, mankind has been almost completely stripped of its spirituality. The coming discoveries of key artifacts around the world will herald the emergence of a new cosmology and a new direction for mankind. People will discover their common ground for existence, a fact the *overlords* do not want you to figure out. At this point wars will end and the work will begin.

The *his-story* of humankind that has been taught to us is totally false. When the insane power hungry priests and their governments colonized the older cultures that included the Mayans, Incas, Tibetans, Indians, Native Americans, Cathars, Celts and Aborigines, they ended choice and freedom. They also ended true knowledge and intergalactic science.

Shamans and Native Elders were told by the continually developing war machine to obey, to worship *God* through a *belief*, when they were directly in touch with the self-empowering reality of this Cosmic mystery. Today's *belief* was once direct knowledge, a sacred cosmic science; the ultimate and absolute truth. To prove my point, consider that the society of this *"belief in God"* is the society that developed poisoned gas, biological war agents, atomic bombs, and who used them on civilian populations.

One of the earliest uses of germ warfare on civilians was during the colonization of America, where the Native Americans, with no resistance to the white mans diseases, were give blankets contaminated with small-pox... Today it is the depleted uranium contamination of Iraq. But they have *belief in God* on their side, and the biggest stockpile of nuclear weapons in the world.

The so-called primitive cultures practiced a very ancient and very precise Cosmology, way beyond what we know today, such was the advancement of their calculations. It was lost to us, but they understood the star and planetary alignments and how this knowledge pertained to life here on planet Earth. The knowledge, called astrology, is finding its way back into the awakening process, as we begin to consider the karmic return of our collective and individual actions.

The ancient's knowledge was beyond the understanding of those carrying swords and now guns. As one part of the world explored and sought to understand the greater force of Cosmos and life, the other part of the world developed more efficient weapons to kill that life. They searched for more effective ways to annihilate larger numbers of peoples... weapons of mass destruction.

Today the regressive hybrid-aliens liquefy humans with light weapons which they deny they helped develop for humans to use; this is the powder keg upon which current society has built its elite status. The foundation of today's ruling elite powers has been built on violence and destruction.

Naturally, those who live by the sword die by the sword. The movement toward mass auto-destruct is not a conscious development of a self-empowered society or species, but, rather, it is the consequence of wrong behavior. The shamanic druidic inter-dimensional contact cultures were the cosmic grand masters of the cosmic chess game. Then along came some bad DNA – which messed up the game. The sage's development was one of practical and valid exploration through knowing, touching and experiencing our world and beyond.

It was a complete insult for these people to be told like children that they had to pay homage to and *believe in* a force that had just killed most of their women, children and elders, mostly through deception and deceit. They were in direct contact with their own spirituality. Unfortunately, many shamans and elders lost their lives, refusing to move from their higher spiritual university to a kindergarten type downgrade of their own high-level intelligent operating system. In this case, death was preferable to slavery.

Today, those people who complain they have a right to their *beliefs* are descendents of the same people who denied the ancient societies the right to their *knowing*. Understanding this is crucially important in order to comprehend why the earth is now facing heart-breaking changes.

Pale blue dot earth has woven humans out of the fibres of her light body. It is not easy for a mother to have to destroy her own children. Do we leave her a choice? The colonizers of planet Earth have destroyed the same wise people who could have guided the world away from these current approaching dangers, not only through education, but by being the guardians and stewards of the planet, of human evolution and its mystery. These high level super shamans and their magi apprentices were able to activate and balance the earth by means of the many sacred magnetic energy centers.

Awareness of the sacred balance was not confined to the earth alone. The magi were able to connect to the centers on all the planets, as well as to the centers on Earth. Mars was one such stop-over. Today, however, humans cannot do this. The ancient races are still not respected by the colonizing military power which is becoming the new ELF frequency power as they develop their super weapons. Not really what we are here for.

The *Ancient Druid* cultures are basically third world countries, underdeveloped and struggling with poverty because they cannot adapt to the ways of greed – the ways of the colonizers. Those who can adapt take on the insane role of more rapidly destroying greater areas of what is left of the natural earth. Now pale blue dot and Cosmos are about to present the Millennium bill for the damages.

Only a race that will lay aside its weapons of mass destruction can begin to move toward self-empowerment and self-knowing – toward peace and then maybe also toward Cosmic wisdom and super consciousness. The star gates will remain closed to those who use violence as a means of control, and evolution will remain a clever idea but not a manifested fact of naturally evolved life. Finding more efficient ways to kill one's own species is not the action of an evolved intelligence, it is the action of a devolved entity.

I observe the earth changes manifesting, from quakes, to shifts to tidal waves, so be safe. Yet, I know there is no reason for fear – if we know how to protect ourselves. However, these first occurrences are also our last chance to change our behavior, mind set and attitude. Now we must go from beliefs to sure knowledge, and learn again how to live. This is a new beginning, a last chance to rebuild a sane planet, while the new grid forms around it like a protective web. It is the emergence into the 5th world.

"Our chiefs are killed... The little children are freezing to death. My people... have no blankets, no food... My heart is sick and sad... I will fight no more forever." – Chief Joseph, of the Nez Perce.

Closer To The Stars

"The limits of the possible can only be defined by going beyond them into the impossible."

~ Arthur C. Clarke

The world is not held together physically, it is held together energetically. The myopic science of the old brain will experience its own earth-changes destruction when this fact becomes the most prevalent truth. Our world and our existence therein are held together through an all pervading omnipresent psychic-ether of cosmic mind, an invisible energy out of which matter is born.

This energy – or sound – is akin to *the force* I introduced you, it is the Mirror of Seeing. It is the unseen force out of which physical matter forms its vibration and therefore its existence. From these invisible realms the psychic network of Cosmic or trans-human over-souls operates in the background of our earthly existence. The illusion is that everything is held in place physically or in three-dimensional space.

There are at least eleven dimensions and eleven universes. In true-reality, all that is, is made of what I know to be a structure of light. In reality it is a vibration of light-sound. Level 11 of our existence is plugged into Cosmos at all times and the master number 11 has light encoded into it. Matter is energy, the actual bonding of a fine light matrix that resonates at a denser frequency to give the illusion, or containment of a limited and restricted material universe. Seeing or hearing this background energy through psychic recall shows me the energetic configuration of events before they happen, so that we can adjust or adapt to the changes at hand.

In the near future, say around 2009-2012, I anticipate world events to create a completely new world, both in the earth climate as well in what we know now as political landscape. Things will change almost overnight. The sea change began when the Western world inherently admitted the strategic superiority of China. Yet, make no mistake, it is not so much a question of what physical or political changes are approaching, but much more a question of the state of the human being crossing into this resonance of total change.

Too much focus is given to the approaching political, physical, and so-called earth changes when really we should preoccupy ourselves with the current state of our minds. In reality, the most important factor is how each individual on earth enters the new time wave or energy matrix. This is a task that begins and ends within you.

It is not a time to focus outside of your own self and blame others, or other forces, for what is now unfolding on earth. The psyche of the future is so entirely different from the psychic matrix we are experiencing, that it is really now a question of how any human being today will cope with those changes. Time will be experienced in quantum ways. Cosmology will be redefined. The secret of life and the ultimate theory of the universe in both macro and micro terms will be discovered, and most importantly, this breakthrough of mankind will be applied in a way that will liberate humanity.

Nothing will be the same once Saturn and Jupiter have passed Pluto in the late Capricorn degrees to move on into

Aquarius. The trans-human structure of everything will be altered forever, and this is a good thing. 2020 is maybe one of the largest scope markers in time for humanity, and beyond that transcendental point we will experience our existence in ways never witnessed before.

My *Star Zen* magic wisdom is to know what to conceal and what to reveal, and when and to whom. The balance between revelation and concealment is what makes this your handbook for your future. The more enlightened, psychic and sensitive ones among us – the few – of course *know* this to be true. Others will take from this what they can and come back later.

It is evident that on an emotional level this time until 2020, and thereafter until 2047 in the Uranus-Pluto opposition, is potentially devastating for those who hang onto the old out of fear. Thought measures reality according to what it thinks it knows – the memories and attachments – and thought fears losing its way in the pathless land of truth. Intelligence is the navigator of truth.

People are panicking without knowing why and you will see in the years ahead many who will literally lose their minds, and with this also their lives. Feelings and emotions are much more intense now than in the past. This appears to be a new kind of *pressure* that has been building up. This energy is so different, that our present evaluation is useless when dealing with these events. Many counsellors are out of their depth now when dealing with the entourage, and so here you have the magi with his Zen of knowing the stars.

Due to the fact that the human psyche has a fixed group mentality approach to life's events, any change

111

hitting upon negative pockets of psychic energy will set up such a charge in those fear based areas that there is no way to exactly predict how masses of people will react in the years to come, other than to state they will turn to complete dysfunctional behavior and this demands some preparation from those who wish to be able to carry on functioning.

This is why I advocate the safe places scenario with all that this entails. Think of it as a network of Rivendell and Helms Deep, or a loose structure of sanctuaries where sanity reigns.

One of the biggest mistakes governments, big business and most scientific or spiritual institutions have made over the last fifty years is to go out of their way to hide from the mass of people that other intelligent life forms exist *out there* in the universe. Even Dr. Edgar Mitchell, the pioneering Apollo 14 astronaut (who presented me with "The Way Of The Explorer" during our stellar Scorpio Buddha full moon conversation in May 2001) states clearly that our future is to go *out there*, to the stars. What he said to me is more valuable than gold: *We will not discover our future, we will create it.*

It is indeed unfortunate that most human beings are content with being deceived and remain asleep in their self-induced slumber of believing that we are the only species in Cosmos. Thank the heavens we are not alone, and let us remember: We came from the stars, and we will return closer to the stars. No one can hide truth from an active, aware, awake and alert mind.

This is a conflict about the nature of reality which I have called the silent war of the magi. The resonance shift is about the approach of planet Earth, in its galactic turn, toward a new power vortex and the magnetic field of a yet to be discovered universe – meeting us within a new realm. This shift is not so much about some unknown planetary body hitting our earth, or our earth experiencing a pole shift. This shift means, and is about irrevocable transformation.

The Aquarian time or age of knowledge indicates that the complete soul group, guided by an invisible realm, comes together and manifest the transformation... together, as one. This is how I see the changes ahead. When I say we, the group of souls, I mean, *all beings* in Cosmos. I address myself also to the extra-terrestrials some of whom are aware that we are thinking together and cooperating also with the planets, stars and galaxies.

We are in this together – all beings must move together and do the right thing. There are many forces of good at work to bring about the transformation of which Krishnamurti – the man from the future – talked about for almost 60 years, or more, during the last Century. The forces of change are multi-dimensional and inter-dimensional. They include human beings, trans-human beings, extra-terrestrial beings, Cosmic beings, and more. They include animals, plants and the creatures of the seas, the earth itself and all planets, everything that is a being and that has light encoded into itself.

We must act like the unified field to leap forward and through the upcoming changes. No elite will play steering committee for what lies ahead. This is a collective endeavour of a kind I know this realm has never been witnessed before. This is the kind of operation that requires team work of the Cosmic kind. No savior, messiah or prophet can see this one through alone. We are to be our own navigators. We are the fine souls we have been waiting for and we are – at many levels – cooperating for the total change to occur.

I know that many human beings of good heart all over this planet are working for the total change now. Some of us know the truth is *closer to the stars*. We are the ones born to walk this cosmic journey. No galactic or other federation will do this for us humans. Remember also, we are not even on our true home planet, as we are somewhat in exile among the fallen ones when we enter this realm, and yet, this is all part of the experiment of being human.

We are ambassadors, time travellers, free to journey and explore creation, free to experience, develop it and be it. There are no "rules" unless we create them. The infinite cannot be contained by any laws created by man. It is time we stopped limiting ourselves and return now to our journey of oneness and wholeness.

It appears that the resonance of the changes approaching us is not as important as our inner ability to meet this force and attain enlightenment. How will we enter enlightenment if not through our own existence? The human essence is to be aware of what is happening by using what we discussed in this book – the Mirror of Seeing and the Zen of Magic – at the 11th Level, with the 8th brain functioning harmonically. The purpose is to make this Cosmos a peaceful and interesting place to live and evolve in.

The psychic recall functions of experiencing life at a higher or 11th level can be developed by each and every human being who is willing to learn anew, and who is able to cast aside the conditioning of the hierarchical mind. Each harmonic is light encoded, light and sound being of one source. These light-sound harmonics have to be understood and also used as part of a new magnetic *free-energy* source.

The vast majority of mankind progresses through this business called life with eighty percent, or more, of the brain asleep. It is the larger – yet sleeping – resource of our trans-human soul-mind that is now activated and expanding to reconnect with the Cosmic mind via the intelligent universe or star guidance, to create a new and better reality. This doorway is given to us by the invisible world at the moment we enter this realm of seeing the *Zen of Stars*. It is when we become truly one with this guidance that we begin to enter a state resembling enlightenment. Planet Earth wants an enlightened population. The sun is now awakening that resonant light-sound within us.

Trends Forecasts & Nostradamus

"What is sacred? That can only be understood, or happen, when there is complete freedom from fear, from sorrow and when there is this sense of love and compassion with its own intelligence. Then, when the mind is utterly still, that which is sacred can take place."

~ J. Krishnamurti

Following are a series of trends and events leading to the eventual collapse of the *New World Order*, and a shift into a new time, bringing about the much expected *passage to peace*. Among my past predictions made in 1999 in seminars and published in writings since that time, were the 2000 ballot recount, the Bush presidency, a number of leader's elections and re-elections, and many others documented on my website, passage11.com.

With the passage of Uranus, ruler of Aquarius, into the sign of Pisces, my work is transforming now from point to point predictions into the formulation of mega-trends up to the Pluto-in-Pisces versus Uranus-in-Virgo opposition alignment of 2047.

The one valid question to ask at this point is: How does it fit together with what we know and with what we have yet to find out?

I don't purport to give conclusive answers, but rather, I invite you to ask questions and draw your own conclusions as to actions to be taken or avoided by using your own independent thought.

I have stressed before that each person should develop their own inner psychic recall ability. Also, if we humans do not see ourselves as a part of humanity, then we are essentially the enemy of humanity. Human mindset has fundamentally not changed in over two millennia, but is due for a massive upgrade in knowledge.

Based on study of the parallels between planetary alignments and historical events dating back 13,000 years – half a Great Year cycle – spanning half a dozen ages, the present world is due for a decisive shift in energy, consciousness and awareness levels. A crack in the wall of perceptions is inevitable, as seen in the discovery of a new form of cosmology.

The age of deception and "double speak" – prophesied by George Orwell in 1948 in his work, "1984" – will last for a few more years. The world will *not* end in 2012. Many theories and pieces of the puzzle are missing, and when spiritually sensed and intuitively understood, the alignment of planetary calculations can be a useful tool to interpret new developments.

It is important to develop more silence and serenity around ourselves, and also cultivate longer attention spans, in order to study and learn. Each new understanding is necessary for comprehending where we came from and where we are headed. The end of 2007 and beginning of 2008 is the time in which I see a manifestation of the great inner and outer shift.

A strong psychic and parasitic force, designed to enslave us through the mechanism of belief systems based on falsehoods, is about to find itself face to face with awake human beings. This in itself will bring about much needed change in the way we go about our lives.

The now hidden scientific work of Nikola Tesla, *The Master of Lightning*, was based on his knowledge of a background ether, creating movement in the universe. He predicted that, as we find a way to transmit this invisible force technologically, we will have *Free Energy*. After fourteen years of exploration and research, mathematician R. G. Moon, author of *The Vortex Theory*, has proven scientifically that *Ether* or a background force is moving the universe. As this theory is developed, the outcome will indeed be free-energy and anti-gravity technology with all its positive consequences.

The world will go through serious changes during the next 8-year solar cycle and 5 Venus revolutions from June 2004 to June 2012, which to the Mayans was the completion of the last great cycle. Beyond this cycle, in the upheaval of energies, the *New World Order* will perish and the global economy will be transformed. After a ten-year period of conflict, people will recreate a system of living in liberty, if they truly understand what is at stake, i.e., the survival of the human spiritual light energy shell.

The use of reverse engineering technology, genetically modified foods, electro-magnetic wars, and attempts by shadow governments to rule the earth will fail when enlightened new generations opt out of the system and learn to co-create with nature. Defending individual free will, and taking responsibility for their choices, these enlightened individuals will object to the direction currently taken by the political powers now in place. More important, they will take power into their own hands to change the projected course of events.

118

After studying Nostradamus in his native language for many years, and comparing the French seer's verses to other prophetic pieces of the puzzle, the question to ask is will the world's elite will have anywhere left to hide; for the truth will be revealed. I do not discount the presence of "friendly aliens" who are trying to make their presence known to humanity in order to warn people about the current world order power struggle.

I advocate seeking intelligent contact with extra-terrestrials. It is important that we see through the deceptions that include man-made UFO's, the new reverse engineered toys of the secret government.

Weather manipulation and HAARP mind control, as well as Montauk and the Philadelphia-Phoenix experiments will backfire on those who desire to use the power of nature for negative based control of human populations and the earth. The human spirit will eventually find freedom. Ultimately these control strategies will turn on and devour the forces using them.

I foresee a period of a decade-long struggle for truth. This includes earth changes, severe tornadoes and increasing eruption of volcanoes world-wide, until people throughout the world become aware of what is being engineered and ultimately rebel. This will require an understanding of what is at stake – the survival of humankind. The learning curve will be steep and cost many lives.

In 2002 I predicted the exact date of invasion into Iraq and that this dirty-war would last many years. I predicted Iran would temporarily be brought back under a puppet government and earthquakes would soon start happening in places pin-pointed via the tool of Astro-Cartography. All of this has happened as predicted; and much more disruption is in store. I have selected two key quotes from Nostradamus – for they have astrologically discernable meaning:

"The man from London – appointed by the government of the rich souls of America. Over the island, which is locking horns – the era marked by freezing. The king of the Baptists makes a great mistake toward the Antichrist. Because they invest everything in the leprous souls."

* * *

"The black one dissolves big government so he can rule. The power of the weapon determines the border of the countries. Time of open air and the duke is happy. Destruction is carried by the new sounds – day of the stupid one."

Uncovering the truth is of paramount importance in my vision of future world peace. It is too early to come to conclusions, because I feel many prophecies of the past are coded time-capsules to be understood by our leading minds only when the conditions are ripe for change. The issue of free will and choice is paramount in the changes that will continue to occur.

Secret experiments will create global disasters. This in turn will make people demand the truth from their governments. The global public outcry will be so tremendous no politician will be able to ignore it. The dangerous trend is toward a one world government, unless populations unite to globally respect nature, the earth and

Cosmos. Any attempt to create a non-elected self serving secret government will ultimately fail. People involved in world domination and mind control projects will begin to talk about what has been done, because of the disastrous effect on the weather and on human beings alike. The whistle-blowers on government black-operations are being warned by higher dimensional light forces to come clean.

In the star charts, I see a period of changes up to 2008. Then we will choose to decide whether we want to be slaves to a deceptive, corrupt and suppressive system, foregoing our God-given right to happiness and fulfillment of optimum human potential – or whether we value our freedom enough to opt out and even oppose these tyrants globally. By necessity, I predict people will opt out.

Having chosen freedom, people will ideally work together to establish a peaceful world after the inevitable and devastating earth changes. Only a few will physically survive these changes. Future alignment of planets over the fixed star of the Southern Cross mirrors many societal trends. The obvious one is that in order to see the Southern Cross rise over Mecca, Medina and Jerusalem, the globe will have to tilt slightly; hence, a pole or earth shift is probable.

Discovery of new technologies is another potential parallel, mirrored by simultaneous future planetary alignments. So is the possible peaceful rise of the "wisdom of the East" following a return to the ancient doctrines of Atlantis – moving toward a pure scientific or mystical theory of reality in the form of a holistic Cosmic view.

I confirm via my own intuition some of Edgar Cayce's views, by which the true history of humankind will be known after the re-discovery of ancient scrolls in the

Vatican and at key sacred sites. We will also discover other documents, tablets and manuscripts hidden in the Hall of Records beneath the Sphinx.

Why is a wall being built around the lion star gate of the Great Pyramid? And why does the New World Order attempt to engineer humanly contrived confusing prophecies?

The Sphinx stands for a transition or blending of the age of Virgo to the age of Leo. I have confirmed via modern astrophysics calculations, changes in the light frequency around the earth, as seen 13,000 years ago, and foretold by the Mayans. These changes will make it impossible for the present deception to continue. People involved in cover-ups will come forward, because the evolutionary process will begin to accelerate with the emerging of a better understanding for a peaceful world. I foresee the passage to peace to be completed by 2047.

The quickening of time can be seen in the context of *The Vortex Theory*, by which cosmic motion exists and time does not. I continue to collect data and opinions from leading scientists and mystics from all over the world, and then I compare the pieces of the puzzle. I encourage others to do the same, without jumping to conclusions.

I would like to stress that no one has a definitive answer concerning matters pertaining to spiritual, technological, political and societal trends. This is because of the very definition of the perpetual change and motion of the planets. As foreseeable as they are, if we can predict the outcome, we can change it. The best I can do is bring together as many pieces of the puzzle as possible and let them stand the test of time and the scrutiny of my field of expertise. This area of research and knowledge is a new yet ancient form of Cosmic viewing that combines the spiritual and mundane by delivering astrophysical facts paralleled to world affairs in recorded human history.

Ongoing observation of the Southern Cross in mid Scorpio during a zone spanning a few weeks in September of 2013 points clearly to a future *beyond* 2012. The simultaneous passage of the North Node and Saturn (time marker alignment) in 2013 emphasizes this fact.

Once the anticipated upheaval on earth, the wars, civil unrest and massive topographical changes are history, will there be a light-age of awareness and the ending of wars on earth? Will there be the emergence of a Cosmic consciousness?

Planetary alignments in the nine year passage of 2012-2021 point to a high probability, with an emphasis on the shamanic idea that, *we are the ones we have been waiting for*. In other words we are not waiting for superior all powerful gurus, saviors, prophets, anointed ones, ascended masters, avatars, and gods of all sorts. Do not give your power away to anyone or to anything!

Evidence of a secret government involvement with aliens will soon be public knowledge. The wars in the Middle East are an attempt to gain control over a lost technology that is being kept secret from the world.

The *New World Order* forces will try to create a uniform society in which every human being will have an electronic chip implant or something similar. Implanting of this electronic-chip will also be compulsory in hospitals and mandatory at the birth of each child. I leave it up to you to decide where your child should be born, in that case. The emergence of a world religious leader will be engineered via projected holograms and other deceptive techniques. Will it fool people? Acceptance of this holographic messiah will mean the loss of individual freedom. Which of us will see through the deception and reject it?

Attempts will be made to bring down the Internet, but new security software will be created to keep the World Wide Web alive. We will find a new way of transmitting Internet signals. The Internet will be like the web of life described by the great Chief Seattle.

The world wide consumer economy as it currently exists will fall apart and the new world order insanity will perish in this upheaval. The secret spraying of chemical toxins in the skies (Chemtrails) will result in burning and the downpour of acid polluted rain. This will destroy and damage trees and plants. Extreme numbers of people in the cities will become ill and there will be a rejection of any form of dangerous drugs, as people rediscover the power of Nature, of natural herbs and healing plants.

Tom Brown, Jr. describes Stalking Wolf's Vision of, *The Red Sky Prophecy*: "Grandfather said that there was not [a set] future, only possible futures. The 'now' was like the palm of a hand, with each finger being the possible future, and, as always, one of the futures was always the most powerful, the way that the main course of events would surely take us. Thus his predictions were of the possible future, which meant that he always left a choice."

He writes in Grandfather's own words how we have to change the way we live and begin to make the right choices, as the only way to alter the future. Grandfather says to Tom Brown, "If a man could make the right choices, then he could significantly alter the course of the possible future. No man, then, should feel insignificant, for it only takes one man to alter the consciousness of mankind through the Spirit-that-moves-in-all-things. In essence, one thought influences another, then another, until the thought is made manifest throughout all of Creation. It is the same thought, the same force, that causes an entire flock of birds to change course, as the flock then has one mind." [Tom Brown, Jr. *Night of the Red Sky: The Prophetic Vision of Grandfather* – Nexus Magazine, December, 1999]

The truth about Atlantis will re-surface. Evidence of advanced civilizations will be discovered under the sea and under the Sphinx. New islands will rise above the waves, revealing advanced technology from the distant past.

The true history of humankind will be known, where we really came from and why we are here. We will find out where we have come from, as news about pyramids on Mars becomes public knowledge. Science and mysticism will merge back into a Cosmic consciousness of spiritual science, the science of nature, the science of sound and light. We will finally learn to work together with the intelligence of the earth, stars and planets.

People will migrate in the form of spirals following the mathematical figure Phi, [1.6] which is the *Vortex Spiral* or a key function of life. Now you can enter the star gate with courage, via your heart, the lion heart, which connects to the lion gate via knowledge. You must not believe that you are, you must *know* that you are. "I think; therefore I am" is old hat teaching by Descartes, yet not untrue.

"I know; therefore I am" is the true age of Aquarius type age of understanding, or the age of knowing. The age of beliefs that started 2,000 years ago has now ended. The fish has been spilled onto land in the urn carried by the water bearer. Welcome to the age of knowledge. Welcome to the passage to peace, characterized by the passage of the three great planetary avatars of change through the sign of Pisces and compassion until 2060: Uranus, Neptune, and Pluto, in that order. Uranus is in Pisces until 2011, Neptune is in Pisces from 2011-2025; and Pluto is in Pisces after 2043...Future warfare is indeed a war of ideas, and the information is the war itself.

Transitions of Beings

"Drink me."

~ Lewis Carroll, Alice In Wonderland

From 2007 to 2017 we will live through what I have termed the plutonian decade of planet Pluto tearing apart the structures of Capricorn. We will witness total chaos, violence and implosion of the larger cities. Pluto is known as the breaker of resistance.

Many will leave the cities. This is the shaman's process of purification that the Hopi Indians have warned us about for decades. Cities without power will become death traps in which crime lords ruling the local neighborhoods will do as they please. Police will be unable to maintain order, and the military will be called in as a necessity. However, they are spread too thin across the world to be able to stop the violence and crime.

These warlords will reign for a period of about ten years; some populations will be put under curfews, or a state of Martial Law. For anyone wishing a future, it is now time to make alternative plans.

Employment will diminish because much of the former labor force for industry will come from prison reform camps. This will be hailed as a miraculous plan for socially reprogramming the exploding numbers of criminals who must be kept in the penal systems. At first it certainly will seem to be the answer to the problem of the west losing its manufacturing to China. However, when the working classes no longer have a source of income, they will swell up as an angry mass and end up interred in concentration camps.

Mary Summer Rain viewed the coming earth changes with her Native American teacher, *No-Eyes*. This is also what I tend to see in my own visions. I repeat that I see most of the flat lands of the old world under water.

She writes, *"I followed her above the area of the United States. I barely recognized it... Everything was tilted down to the right. Alaska was now the tip of North America. Mexico wasn't south anymore, but rather west. New York was only partly visible. On a closer inspection, we found that all of North America's east and west coasts were gone. Florida appeared to be ripped entirely off the continent. The major fault lines had cracked.*

"The San Andreas ripped through the land like some giant tearing a thin piece of paper. The torn shred drifted out into the churning ocean. The waters of the world literally swished back in one huge movement, paused and came surging back to a new level of balance. This great movement washed over hundreds of islands. Hawaii was gone completely. Borneo, Sumatra, Philippines, Japan, Cuba, United Kingdom – all vanished within the blink of an eye!" [Mary Summer Rain – *Phoenix Rising: No-Eyes Vision Of The Changes To Come*]

The *Q'ero* (Long-haired ones), the last of the Incas, recently revealed their prophecies of the *End of Time* to Alberto Villoldo. The *Q'ero* are awaiting the next *Pachacuti* (He who transforms the Earth), and expect it to be the end of the world as we know it. The signs of upheaval have begun, and will last many years (2003-2013). I say it is not the end.

A new humanity will emerge from the chaos. The prophecies announce the beginning of a new "millennium of gold," and they speak of "a rip in the fabric of time," through which will come luminous beings of light. The signs of the times include the drying-up of high mountain *cochas* (lagoons), the near-extinction of the condor, and great solar heat. Afterward, we shall emerge into the 5th Sun.

I disagree with the *End of Times* terminology but my calculations match their narrative. Structures of global economies and banking systems will certainly be destroyed. By 2007 there is no way to uphold the lie that things are all right, as we see literally millions of hard working people in the G8 nations lose everything to the world's economic disasters engineered by the corporate banking piracy.

Insurance companies will fail because of increasing natural catastrophes, wars and weather calamities. Industries will implode due to inflation; health facilities will also fail. With no cures for major toxic diseases, there will be no way to deal with the huge numbers of dying and sick in the hospitals. At least half the population will not survive. Shamans tell me it is more likely that 80 percent of the people will die during these disasters.

A global plan to depopulate the planet will actively go into effect, eventually causing massive revolts and the tearing down of existing political systems. The situation in North America and parts of Europe will evolve into a mutiny of the military forces who divide into opposing factions, and even form gangs. Some will be upholding the agendas of the prison planet police or the *united-nations*; others will remain loyal to the old military régimes of the last century. Local militias will defend their enclaves from rioting gangs, and still another military type inner army will emerge as a new Western States Defense.

The US will segregate into a few distinct regions: East, West, Mid-West, South, and Central. The South will definitely secede and fend for itself. Florida will be a "give-away" to the Cuban Caribbean pirates. California will go to the Chinese. The Western States, which are not entirely reliant upon foreign oil and exchange in the European markets, will stand as the last bastion of the decrepit "American Dream" based on the spiritual ideals set forth by the misguided founding fathers who stole this lands from the indigenous Native Tribes.

Eventually China will overrun these states. Racial violence, riots, corruption, lack of fresh water and loss of electricity as well as social depravity and global crime will turn cities into virtual war camps. The films *Blade Runner*, *Water World*, *Post Man*, and *Planet of the Apes* or *The Day After Tomorrow* are relatively accurate depictions of what can happen. The governments will hide underground in the mountain areas.

This chaos will be experienced world-wide. Saudi Arabia will soon go up in smoke. A few limited nuclear exchanges as well as other humanly contrived incidents will occur. The European Union will create a new alliance between Western Canada and the corporate overlords under what I term a newly reformed Western Hemisphere Pact. This will consist of what will be left over of the Mid-Western US, Mexico, what's left of Japan and some expatriated Chinese moguls living in what was once called Taiwan and Hong Kong. Western Canada will feature prominently in this new alliance pact.

Globally this will result in three blocks, as stated by Orwell in *1984*: the Western Hemisphere Pact, the European dominated Euro-Russia pact, and the New Global Alliance forming a third world; holding the majority of the global industrial resources under Chinese rule and the rest of the world in non-aligned and tribal-based communities.

Asia and the western hemisphere pact are already headed toward future integration. As this unfolds, we witness an increasing destructive behavior of weather patterns that may derail the pact's alliances because traffic will be via sea, once air travel ceases to be safe. The three spheres of influence will realign until a tripartite peace deal is established. We will therefore remain in a multi-polar world for a while.

Electrical and cosmic sun storms will occur, with lightning causing fires due to unusually long droughts. A series of 300 mph tornadoes will whip through the country, one after the other and dust storms will also blow across the lands. Farms and ranches will lie fallow. Rain and snowstorms in sunlight, without any clouds in the skies, will be prevalent in the mountains; as well as snow, blizzards and orange-size hailstones in the heat of summer. I cannot say for sure which areas of the world will be touched, but my impression is that Europe and the US will be hit hard.

Firestorms will rage for years, creating fire tornadoes. Volcanoes will erupt across the globe, and under the oceans, causing lakes of burning lava and destroying entire cities in some places. It will be dangerous to live in forested regions, as areas will continue to smolder with only the winter or summer snows to slow down the advances of total destruction.

Living at sea still seems to be the best option, although the rapidly rising waters will claim hundreds of miles of land in the coastal areas. Water communities will have to defend against pirates and increasing lawlessness at sea. I saw airplane disasters because the planes will be hitting hard walls of water with no warning in sight. Thus, air traffic will greatly diminish in this time, due to uncertainty and fear.

Ice sheets will fall from the dark sky, first like a mist encompassing everything until the fog is so thick, it will appear in satellite pictures as grey blankets. That is, if satellites are still up and running. One of the visions I saw is the demolition of the sky structures built by aliens and the secret government; so I feel many satellites will come crashing down to earth, together with red rain from the clouds. This will cause night during daytime. Death will follow quickly because the air will burn the lungs and eyes will be blinded from the damage. We will have to be prepared to stay inside for long periods of time.

Fireballs and meteorites with the consistency of razor sharp metal, will fall from the skies and cause more fires and destruction. Giant spouts will appear in the oceans and later also in the greater lakes. These huge funnels will spew things up and out at tremendous heights and distances, and at enormous speeds.

Hurricanes as we are seeing them now are miniscule compared to what will occur. I was told we will see appearances of strange rainbows at night, and the sky will glow red by day. People will fear the end of times, although others will interpret this as a sign of divine intervention. This will not be the *end*, my friend, but the beginning of new times. Just be safe!

There will be intermittent still phases with no movement of air, and pressure felt in the ears, which can cause disorientation. Many people will lose their sense of balance because of this disconnection from the life force energies. They will lie upon the ground as if in seizures. I see this as the "ripping of the time fabric effect" which we will witness for a while, probably between 2007 and 2012.

Cosmic radiation storms will stop normal life for a while. Businesses, schools, public airlines and sea traffic will be shut down, and the public will be told when it is safe to leave their homes. Learn how to live without electricity. Contamination of crops due to these storms will cause havoc in the world's economies for many years. In 2003, Europe, France and Italy especially, experienced thousands of deaths due to weather anomalies.

All of this havoc will cause serious spiritual re-evaluation of human beings outside the system. They will return to a simpler life style, closer to nature. Spiritually, Christians as well as Muslims will orchestrate mind control and deception to maintain control. One individual will rise, proclaiming loyalty to the patriarchal dominance of old religions with their archaic perceptions of "beliefs." Some will be taken in by his apparent miracles in the western hemisphere; others will see through the deceit and realize the true political agendas secreted behind the make-believe ideologies of a new world religion.

Many wise children, by 2007 in their late teens and guided by enlightened beings and ancient souls, will create their own miracles, to reverse some of the effects of the devastation and death on the planet. They will facilitate the realization of our true spiritual heritage and

133

inheritance from the stars. That is the moment, after 2012, when we will look into space for new beginnings and new adventures.

As discussed earlier, this narrative is heightened by the obvious presence and revelation of a long-term relationship and hidden agenda with the existing alien beings already inhabiting the earth, below ground and among us. Many of these groups will start to openly interact with global governments in their attempts to create a political peace among the warring factions inside the three world pact alliances, during the failed takeover of the world.

Programming through mass propaganda and enforced rule of law, government deception and mind control will increase in this time, culminating in an electronic era world wide *Anti-Christ* deception. The anti-magical rule of new age faith-type beliefs will crumble from within, after a last ditch attempt at global oppression and mental dominance. Watch out for fake prophets and false leaders!

This epoch lasting from about 2005 to 2015, will be reminiscent of World War II; alien technology acquired by the US after the defeat of Nazi Germany will be used against protesting populations. This is why I describe it in detail and suggest that we prepare for it. Eventually, toward 2017, the seemingly invincible dark forces will turn on each other and annihilate themselves. The Draco-snake is about to eat its tail as a sign of the silent rebellion everywhere in the "system's world" which has succumbed to materialism and fatality.

As the spiritual cross-culture ties and social fibers of the Western Hemisphere populations start to merge with Oriental soul practices of tolerance, the open application of multiple spiritual philosophies in clans and migrating tribes will emerge as a dominant reality. People will integrate rapidly, clarifying aspects of their own version of a new world. Thus, the true Buddha, Krishna, or Christ Consciousness will blossom with advanced scientific knowledge among those who will survive the changes. And then the reawakening of the middle way will have begun, via knowing – but not as a belief system.

The survivors will use a new road map that will definitely include astrophysics. Based upon the commonality and blending of many truths which exist within the Cosmic and ancient native educated cultures prevalent within migratory peoples, we have already formed an infrastructure of a new understanding.

This faith via direct knowledge will include the wisdom and teachings of almost every culture and faith that currently exist on the globe. It will experience rapid growth among the newly developing tribes that will emerge during migrations. These will begin in the western hemisphere due to the acceptance of gender equality and honoring of the divine feminine, which is almost totally lacking in the patriarchal European régime.

This science-based faith, or direct knowing, is *not* the one world religion that will be pushed on the people via deceptions, such as holographic messianic images appearing in the sky. The new faith is an underground type of knowledge that has already begun to surface and is being discussed among the awakened few. The survivors will discover they need each other.

Much of Europe and the US will perish in the flames of self-inflicted devastation. To my European friends and readers, I want to tell you, it is time to transform the doomed lands of the Roman Empire. I feel that an enlarged waterway or new ocean will split the Americas. The quakes in Guatemala show the plates will rip at that fragile bottleneck.

This passage and zone of changes is the phase of the "Transitions of Beings" as I term it or the crossing of times... not end times, however. I know it is anchored in deep understanding of the fact that we must make this crossing or ascension in a spiritual and practical way while still functioning here physically. Spiritual behavior is not a vacation retreat, as some of my misguided friends used to believe; nor is it a hobby for spare time. On the contrary: the path with the guidance must always be followed and applied.

Although some of the survivors of the evil wind spirits will understand this message only a short time before the catastrophic events occur, many others will be learning these teachings out of sheer necessity for survival in communities where millions of homeless women and children will already have been forced to flee from cities that have experienced military suppression or internment in prison camps.

These are the rewards for those who do not wish to participate in the system for salvation, as postulated by the corporate greed of the electronic control chip one world government. *"Do not take the mark of the Beast"* will become a slogan for the underground. The vast silent majority of humanity will join a tacit resurrection of the human spirit and cosmic wisdom already beginning to blossom in rural areas throughout the world. It is also rapidly finding its way to urban areas via the Internet, and books like this.

The most obvious concern for people would be chip-implant identity theft and duplication. Recently news is breaking through that the few people who have been injected with the glass capsule containing the electronic chip have a security problem. Research has shown how easy it is to clone a chip implanted in a person's arm and program a new chip with the same number.

How do you prove that your chip-code got stolen and cloned? What do you do about the bills appearing in your bank account? Or what do you do when criminal organizations are using your identity to break the law? How do you prove it? How do you prove it wasn't you?

Many well-read people misunderstand the so-called New World Order. Some of my friends and I know the Old World Order that originated in Europe will be established world-wide by using the established forces seeking to hold onto their power structures. They will install hierarchy based governments and ideology just as we witnessed in Iraq – and which I warned against, in my December 2002 open letter to Bush and Blair.

More traps will be sprung. The new US Constitution already exists as a hidden United Nations secret document, a one world charter. It will reveal itself as a deadly sting because it is based on taking away all human freedoms. At that point people will start to wake up.

The chaos and human confusion I foresee arise from the desire for world domination. The open and secret wars in the Middle East are taking place now. The secret wars are the result of a carefully hidden power struggle by leaders who do not want people to make their own decisions. Those who planned the takeover follow a written plan laid out for them hundreds of years ago. It is this kind of behavior that caused the brutal takeover of the original Celtic cultures of Europe. The colonizers think they can use this unstable structure to rule the world. The tragedy is they are incapable of ruling anything. If you lift the veil of deception, you can see their one-way road to doom taking form.

The revival of a true survival science spirituality without bondage or false beliefs and with cross-cultural tolerance surfacing in the Western Hemisphere will be one of the empowering forces that will cause unusual events, or miracles to manifest. Eventually the light will overcome the dark restricted perceptions of past world régimes. This is a magical spiritual war of Cosmic life versus alien technology in misguided hands. It is one of consciousness and awareness against dying elites who were able to establish their power-base, because people did not pay attention.

This coming head to head amounts more to a *Spiritual Special Forces* commando type protection of the earth, and not a chanting tea party in an Indian ashram. These newly revived wars of the heavens surrounding the régimes of the alien *Anti-Christ* will be devastating to the socio-economic systems of the Northern Hemisphere.

The Western Hemisphere will meet an abrupt end as we witness the increased activity of cataclysmic events and begin to experience what, in the last century, was called earth changes – a euphonious word for mayhem. Some advise that it would be best to move as far away as possible from any civilized area. The choice is yours.

Communications will have to be via telepathy. Violent reactions will be experienced from the sun and reflected as increased movements of the tectonic plates. During this time we will see increased interaction between humans and inner-earth creatures as aliens emerge from their hiding places below the surface. Their hidden presence has been too long overlooked and kept secret. Simply because you do not see them, it does not mean they do not exist.

Originally published in the 1930's by a mysterious Dr. Doreal, *The Emerald Tablets of Thoth the Atlantean*, are today considered to be *channelled* material. The writings quickly became an underground sensation, and still are.

Emerald Tablet 8 – The Key of Mystery

Far in the past before Atlantis existed,
men there were who delved into darkness,
using dark magic, calling up beings
from the great deep below us.
Forth came they into this cycle.
Formless were they of another vibration,
existing unseen by the children of earth-men.
Only through blood could they have formed being.
Only through man could they live in the world.

(Translation and Interpretation by Doreal)

Extra-Terrestrial Guidance

"It is change, continuing change, inevitable change, that is the dominant factor in society today. No sensible decision can be made any longer without taking into account not only the world as it is, but the world as it will be."

~ Isaac Asimov

Is decision making synonymous with guidance? Decision making is something I had to learn at the latest by age 22 when in the military. Even in this Swiss army that I served with, I was encouraged to listen to my own inner guidance. Intelligent Cosmos has a way of interacting with me so as to be there in the intelligent moment. This intelligent moment is when all factors come together for a real time intelligent decision.

The planets are a form of extra-terrestrial guidance, and if you can listen carefully enough so is the *Galactic Center*. The spirit ancestors are a form of extra-terrestrial guidance, and as we are from the stars, we are also a form of this guidance. This is having the ability to adapt to a situation and make clear decisions as to how to proceed. Knowing how to act, or when to do nothing and take no action.

"Man is a star bound to a body..." claim the *Emerald Tablets of Thoth The Atlantean*. I would say we are all star based guidance systems, but you would have to be in the guidance system to know. This is why I say you are your own inner light. We can communicate with higher forces, but then we also have to put it into practice. Which is why I said at the beginning, I learned to make my own decisions pretty early on in life.

In my travels around the globe I have had contacts with different types of aliens who are interacting with the earth. Like us humans, they are a mixed bag. Not all are friendly to us, and not all are evil. I have received communications from Reptilian races who are sympathetic to humanity, but who do not like what we are doing to the earth. They are not to be feared. Although some have had more negative experiences, mine have been peaceful. My Nordic Blonds friends are highly advanced.

They have given me many messages of peace and of reconciliation. They wish only good. I am aware of the negative Greys and regressive Reptilians. People I know have had interactions with them and it seems as if the other races protect me from those encounters. Certain Reptilian type beings have at times helped me, just as they help other shamans to survive the earth changes. I have learned to deal with the regressive types by using a psychic force field.

Even the more regressive ones have to help the light workers if they want to change their karmic path. Each of the groups are at different evolutionary levels, but all of them agree on one thing: Make peace, end all wars, and interact through intelligent cooperation. The Nordics are in my opinion the most advanced. It is said that the Pleiadians are equally advanced and peaceful. You could compare the extra-terrestrial races to different nations on the planets, some of them are technologically ahead, and others more spiritually evolved; they work for peace for all. Each extra-terrestrial interaction is different, each contact appearing to carry out different functions.

Above and beyond this extra-terrestrial type traffic, there is a much more mysterious contact – a light that I call *The Guidance*. It is an inner telepathic navigation system of knowing when to do something and when not to. It works globally and only via the psychic star field. This contact is similar to the beings in Doris Lessing's, *Canopus in Argos: Archives;* if you have studied her work you know what I mean. It is also like the force that teaches and guides the invisible Buddha awareness. It is akin to the ancestor guides who teach the Native American shamans. The Guidance works in Zen ways.

This guidance is a network of higher beings of radiant Cosmic light. Impossible to describe in linear or non-psychic ways, for it only operates in the psychic realm. This radiant energy or intelligence is all-powerful, precise and unfailing. Beyond any type of magic, it is what makes Cosmos work. I know this type of guidance is not physical, since it consists of highly evolved luminous beings or invisible forms. It works in mysterious ways with all the planets, species and galaxies faster than thought. It is the sacred force that guided our ancestors in times of high magic, and it is the force that guides some good extra-terrestrial intelligences as well.

These beings are closer to Krishnamurti or Buddha and his teachings. They are architects devoted to the spiritual evolution of each race. They are the source of what I refer to as the *Master of The Light*. This guidance is a form of presence. It is its own power. It can form and trans-form itself and us. Before the guidance can connect with us, it requires our complete attention and openness. These Cosmic guides are beyond the comprehension of the other alien races and their human contactees. No one knows who or what this network of guides is, and only few can even begin to comprehend it. One could say they are the Jesuit order of Cosmos.

The galactic hierarchy tried to create their own kind of *order* as to who or what is in charge of things. Extraterrestrial races belonging to the hierarchy believe in levels, structure, centralization of *power* and in creating order through a spiritual or technological caste system, which is of course something I don't like at all. They wanted to put order into Cosmos by creating certain religious belief systems that would evolve species over time according to their plan by braiding thought patterns into DNA, and such other nonsense. But they are not facing the fact, as Krishnamurti pointed out, that this structured evolution is not happening. We are devolving not evolving.

Everything coming from this hierarchical order is stationary. It is caught in the net of static structures and thus, is not growing. If evolution would have to take place within the ordered structure of the hierarchy, it would already be known and done. As it cannot be known, then their structures are a hindrance to Cosmic evolution and change.

This *Inter-galactic foundation* which without a doubt exists, is not interested in the way my *guidance* works, or in working with these other elemental races. They don't like things that are not controlled according to their system. I was told that Krishnamurti, while teaching and speaking on earth, was also talking to the hierarchy's alien races. He was a messenger to both humanity and aliens, telling them they have to change.

Many of these extra-terrestrial species are much older than humanity and do not pay attention. That is somewhat arrogant because human beings developed much faster and according to another master plan that is Cosmic by nature. They do not understand this plan, and they do not want human beings to bring changes to their billion-year-old static thought structures and systems.

I think that is why many of these species are not happy to have us around – especially because we are going to change everything and catapult their worlds beyond attachment to their stranded hierarchy. This is intended, or part of the plan, according to the guides who have been talking with me. If humanity evolves as is intended, it will bridge the gap between guidance and hierarchy. The hierarchy will have to interact with that mystery and take it into account. Then humanity will become its own master navigator.

I know the hierarchy races are trying to sabotage the subtle changes evolving here on earth. They are not bad beings per se, and they believe they are doing the right thing, because they see us – through the eyes of their hierarchy – as a threat. I guess for them this is not funny, here is this gung-ho type species called humans with potentially fully awakened DNA, ready for entry into Cosmos, and no one knows what we are going to do next. For us it is totally natural and easy to shift and move through all levels and layers like the Silver Surfer in our childhood comics used to do.

Humans are the kind of species who, when given a highly advanced interstellar space craft that is used for transporting goods, technology and services from one part of the galaxy to another would get bored and goof off.

Imagine you place humans behind the controls of an inter-galactic Ferrari and they suddenly start to race each other, to see who can deliver the goods faster. Then, when their friend is winning, the other one manoeuvres in such a way to get an advantage at faster than light speeds. While they are racing each other, the two pilots get the crazed idea to see if they can use the crafts to reach the outer edge of the universe. I guess this is not so funny if you are a hive mind – do what you are told – type alien waiting for the delivery of your supplies.

"Where is Commander St.Clair?"

"Sorry sir, he upgraded the warp drive and shifted to level eleven. The ship's gone. He's no longer in this universe."

As I stated before, such is the human spirit that no one can understand us. Or only few can. The other species generally do not behave like this. If they have an assigned task to do they carry it out, acting in a functional, precise, serviceable and planned way. But a human will swoop down and rescue someone in the middle of a mission. It is what some call the unpredictable human factor. Human beings are the Joker card of the universe, the Fool in the Tarot.

The innocent fool does crazy as well as creative things; and as enlightened beings we will continue to do highly unusual or unpredictable things forever. All this is part of the direct knowledge I gained from the guidance system that has seen me through so far, and I can disclose to you that I was a reluctant student of theirs. It was all new to me at first, until I began to see who we really are.

Today I tend to better understand what the Guidance wants or suggests: Sovereignty. This sovereignty begins with the understanding and application of our birth moment twinned to the moment of death. We are the Fool, balanced precariously between birth (life) and death (re-birth). But in the midst of this seeming birth of chaos we have the ability to navigate this strangeness with fluid intelligence. Obviously the intelligence has to meet the high level of unpredictable behavior that comes with bridging the worlds. The inner guidance navigation system expects the unexpected. The star bound to a body is therefore its own extra-terrestrial light.

"If I want to look at a fool, I only have to look in the mirror." – Seneca.

Celtic Cetacean Delphinus

"We live in two worlds... the world into which we were born, and the otherworld that was born within us. Both may be a blessing or a curse. We choose."

~ Celtic Druid homily

Not much has yet been discovered about the mysterious Pictish and Celtic civilization. Who disappeared as though they magically passed through the master time-portal of infinite design. We read about the Mayan and Native American prophecies and consider them to be highly advanced. I say humans have to go beyond the limited understanding of words to better understand the world that surrounds them.

Let me share the Celtic-Helvetic way of life with you. These people lived very close to the Elves, and the Elves lived very close to the Celts... So who were they and why did they not leave a "message", other than the many strange and fascinating intricate patterns and designs upon stone?

The Picts and the Celtic people were strange in terms of today's world. In fact they could just as well have come from another planet. Today people are awed by the prophecies and predictions of ancient cultures without understanding the whole message. In the Helvetic world, the *ancestors* of the Celtic people still exist and are waiting for this present age to pass before they do anything in terms of what was done before they disappeared. Why did they not leave prophecies?

They had a slightly different view of things. The Helvetic Swiss Celts, the Gauls and the people of Breton are almost one and the same. They did not have to write it because they knew...

First, they knew that anyone in tune with mother earth, with nature and with cosmos would hear and know and see, which is what I do. They also knew that this unfolding or transformation of human consciousness would happen no matter what. Their faith is so complete they know it will happen. Therefore they placed all their power, their wisdom and their psyche into the actual manifestation.

I do not know if I can communicate this in a book? Perhaps I can indicate or signal the immediate need for people to know and be part of the world between worlds. What I cannot do is lead people there. People can only walk there on their own.

I feel it is important simply to pass on the laughter and the sense of humor of our Celtic ancestors. They do not mean to show disrespect for those who are known for their time-capsule warnings, but there is another way to approach this sacred doorway: The way of knowing without knowing.

The ancient Toltecs also left behind no prophecies, no warnings or writings of future light beings saving mankind. Like the Celts and Vikings, they left behind mysterious giant stones, works of art, intricate designs and glyphs that cannot be deciphered through the rational intellect. This means that another part of the mind has to be activated to understand all this, and see things as they are. This tells us that there are other levels. The strands of DNA inside your body do not write linear messages. The

DNA listens to *you* and works directly with *you*. Therein lies your own subtle inner Celtic design.

For those who might misunderstand my own Helvetic sense of humor. I am suggesting that the earth changes are nothing other than the Celtic "time-wheel" Cosmos, turning its infinite course. What humans do, in the meantime, is their own affair. No one can tell you how to live, which is why the Toltecs and the Celts did not tell you. One has to find the way for oneself.

The rational scientific mind measures everything it comes into contact with. In my world I see the Celtic Faerie Shamans dancing on the *quantum* scales, and as no rational based human can see them no one imagines that the measurements are being played around with, as the Leprechaun plays the Faerie harp for the merry dance.

In Celtic Quantum Science the pixies doubted that humans had it right in the first place, as they had noticed no one could perceive the hidden frequencies below zero (as above so below). The pixies generally agreed that humans were showing behavior in the range of -777, but no one wanted to tell the humans the bad news, and as the humans could not see the pixies and faeries anyway, who was going to listen? Buddha did not appear on the scales, because enlightenment is beyond measurement.

It is at this point a large bottle-nosed whale came swimming up the river Thames. Placing the real living crisis of nature at the British parliamentary door. People said the animal was sick, but the whale navigated quite successfully up the river Thames and into the center of London, where its presence endangered the ongoing movement of material goods and the financial human concerns known as shipping.

149

While the materialist fragment of mankind in London was moving Cosmic nature out of the way of the commercial shipping lines, on the other side of the world the first native Indian leader for the past 500 years, was taking spiritual power and carrying out a sacred ceremony in which he and his shaman friends were asking for guidance from the ancestors. There is hope in this simultaneity.

As the bottle-nosed whale swam in front of the Houses of Parliament, Evo Morales the newly elected president of Bolivia was attending an indigenous ceremony at the archaeological remains of the Tiwanaku civilization.

The people of South America's Andean mountains call this time a new solar *Pachakutic*, which means a change in the sun, and a movement of the earth which will bring a new era. According to ancient prophecy, the last *Pachakutic* began with the invasion of the Spanish conquistadores and has come to an end. This is the beginning of a new cycle, one where the world is turned upside down.

I have been thinking a lot about the whale that sadly died after focusing world attention on London, as the new Indian leader was praying and asking his ancestors for guidance. I have been trying to understand the *sign* in this event. I feel it signifies death, the end of the materialist cycle.

Each medicine sign lives out a complete cycle and so we do not fully know what this sign will manifest in the physical 3D world. Obviously this is not the first whale that winds up leaving its natural realm in strange ways. Over the past 14 months, looking back from January 2006, at least 300-500 dolphins and whales have beached themselves all around the world. I have pointed it out when I got news of it, in Tasmania, New Zealand and Australia, or in North Carolina.

For some reason whales and I have a personal bond. The little Cetacean friend who crossed into the ocean of stars while on a barge named *Crossness* has lost nothing and has left us with hope. There is good in everything, if one wants to look deeply enough. Cosmos is simply a larger and infinite ocean, deeper and further away from mankind. What the whale left behind is a seed of hope and goodness. It is much more than a simple warning. We are being shown the way ahead.

Perhaps the whale really felt that London was under water and already its territory, and that the changes that will happen have already begun to manifest energetically even if it might still take a few more years for us to see it. In some way the whale is making us stop to consider how far away from mankind is safest for us to be.

It seems for the wild, free and certainly for the young living beings only certain death awaits inside the poisoned cities of this society. The other pod of whales who stayed together, outside in the wild seas, still have a chance to survive. It is something we should all consider. At the same time it is not only a matter of physical survival. It is not even a matter of seeking the physical means to survive. It is spiritual change that is needed. To travel and be taken care of by the purity of wilderness we have to walk in the spiritual world.

The wild pod of whales who stay as far away from man as possible are uncontaminated by the ways of mankind. They are pure still, in tune with their Cosmic-sonar; seeking the guidance of nature, and having to stay with nature if they are to survive the changes sweeping the earth and its oceans. In a more spiritually aware world the people would have stopped the shipping lines and Thames traffic for a few weeks and observed the whale, praying and asking for *guidance*. In the present world this would require a change of mindset to the spiritual which links the whale mind to the human and cosmic mind.

And yet the one whale in London seems at least to have raised more awareness than the hundreds or thousands that beached in the years before, in equally spectacular ways. The fact that millions of human beings around the world watched this one whale helped to pass a message. The event brought the situation to humanity and to the doorstep of the powers that be, right to the steps of the seat of power of those causing the problem. This was and is a phenomenon, archetypal almost. Similar to the powerful spiritual forces the Atlanteans interacted with. On the most basic level the whale is saying: This will all come back to you.

The spirit of the whale came to my cosmic-window and said to me that sometimes a being incarnates into a form to carry out a specific task. They come to the earth only to do that one thing and nothing else. Taking it further: Who was London created by? The Romans, then it was later developed by the Anglo-Saxons. This sign would imply their end too. People cannot pretend anymore. The whale swam up the river and said: "Here I am. Look at what you are doing to us."

The death of this creature sends sparks of consciousness out into the minds and hearts of everyone around the world. The animal makes a sacrifice. It sacrifices its own life for the survival of its species. Very soon you are going to see special humans like that whale, and you are going to have to keep them away from the old society during the period of transition. Either these beings will suffocate like the whale in the poison of materialism, or they will survive *in the wild*. What is the wild? It is the inner landscape of purity, spirit and love

In *Lord of the Rings,* Tolkien describes the material wastelands destroying the earth as the Land of Mordor, surrounding the Dark Tower of Sauron. Only death and destruction spreads from the dark tower. No living thing grows there, only maliciousness and fear. Sauron's dominion meant death to the free world, to animals and trees, humans and elves, dwarves and hobbits. Nothing would survive his victory if the free world could not unite to defeat him. Even those enslaved would become dark and walk as the living dead.

Sauron's power would consume all of life on Middle Earth. But his power was dependant on the ring. The one ring that would rule all the world. It is clear that Sauron had to externalize his power in order to carry out this world domination. This is why he put most of his power into the ring. Destroy the ring and Sauron's power would be gone and the tower would fall. Middle Earth would survive.

Does the death of the whale signal the death of the forces opposed to nature? This medicine sign has not ended with the death of the whale, it has only begun. That is the strange thing about signs from Cosmos. The sign is a real and living event. It grows in a living way, bridging the invisible world of spirit with the physical world of men. Some wise souls suggest that the Cetaceans sing our world into existence. I would also suggest that the whales and dolphins are cosmic beings living in both worlds.

What would happen if I were to ask *The Master of Light* what he thinks about this sign? This rarest synchronicity of the sun and Chiron in the first degrees of Aquarius with the tribal leader in the high Andean mountains taking spiritual power, and the innocent whale from the deep surfacing next to the tower of Sauron?

The Master of Light points me to the misty mountains above our lake – in a gesture to go find the *Cosmic Contact Center*, using my navigational skills while listening to the Helvetic elves who have guided me with great care all my life, especially when I was navigating in foreign and unfriendly territory.

He would say we are all Cosmic beings living between the realms of the real and surreal. He would point out that we are here to bridge the worlds. Perhaps this is what our whale friend was also saying to all of us, in its own way.

"Someday perhaps the inner light will shine forth from us, and then we'll need no other light." – Johann Wolfgang von Goethe.

The Special Ones

"But there is a power which is in no way related to that power which is evil. This power is not to be bought through sacrifice, virtue, good works and beliefs, nor is it to be bought through worship, prayers and self-denying or self-destructive meditations. All effort to become or to be must wholly, naturally, cease. Only then that power which is not evil, can be."

~ J. Krishnamurti; in his *Notebook*
(July 13th, 1961 Gstaad)

In 1986, when I was 27 years old, it was difficult to discuss astrology intelligently – when in fact to the beings from the stars this occupation involving the mathematics of time is as basic as calculus is for kiddies. I had barely heard of Krishnamurti at that time; I was still absorbed in my own *occult* quest, seeking the hidden knowledge, only to find out it was all full of incongruities.

Later, in 2002, a friend and advisor introduced me to the work of this Indian teacher. What I share with you in this book is some of what I know, and I hope it is of some help to you. I also trust it motivates you to find your own truth and to discover or know what right action is about.

I am not telling anyone how to live, because in fact it is not my problem. Throughout history, sages have made a point of not delivering instructions. Those who are caught in beliefs often feel it their duty to deliver life instructions to the rest of us, based on their beliefs. Some of these individuals also like to accuse others of being part of the *dark force* if they choose not to subscribe to their beliefs.

At some point in the future, I really have no idea how fast or slow, in reality the following will take place. This will unfold because humanity took too many wrong turns, the last of which was the one I warned about in 1999.

Probably around 2009, with Capricorn Pluto when the world will change, some strange yet highly talented children will appear. These children will have advanced knowledge and no beliefs. They will have been born with Pluto in Sagittarius, the ultimate reflection of truth in the obsidian mirror.

These births are *timeless*, in nature, as they belong to no system, no authority and no religious institutions. As a result the special ones will face the same blank walls I faced when I began to tell people what I saw and what I knew.

These kids will be so advanced; it will make our heads spin. I might as well say they will be extra-terrestrial like. They will be extremely developed with skills reminiscent of Atlantis, Mu or Lemuria. They will point to the future and yet seem to come from ancient times with knowledge beyond what people will be able to cope with. They will not tolerate belief systems, nor will they serve any *master*. Most important of all, they will speak the truth. They will know about the catastrophic change that defeated the Atlanteans and also what happened in Cosmos at that time.

These trans-migrational beings come in with visions, foreknowledge and memories intact, spanning over thousands of years. Unlike the present society, they will use and protect their imagination in order to survive. They will have been born when Neptune and Uranus were in Capricorn, which is not for the faint-hearted. Spirituality will be useful or it will die.

In November 2005 news of a 15 year old boy sitting under an ancient pipal tree, who meditates like a Buddha, spread across the world, from Nepal. For six months he had meditated, taking no food, and so the people who heard this began to call him the reincarnation of the Buddha. Ram Bahadur Bamjan has said, "I don't have the Buddha's energy."

He was born May 9, 1990 on the Buddha full moon conjunct Pluto in Scorpio, with Neptune, Uranus and Saturn in Capricorn. Of her son, Maya Devi said: *"Despite the fact that he was born into a Tamang family my middle son was different as he would not eat meat and neither drink alcohol. He never fought with anyone and loved to be alone."* At the time of publishing this book, January 2006, Ram Bahadur has continued to meditate without food or water for the past eight months.

As reported from a Nepalese translation: *"The chilling cold waves have taken dozens of lives in the Terai region of Nepal in last several days. But Ram Bahadur Bomjan hasn't been effected at all. Even in this season of chill, he is continuously meditating wearing just a single piece of dhoti..."* Priest Prem Lama said, *"The biting cold has had no effect whatsoever on him so far."* The point I want to make is that it is not the special ones who are strange, but the ordinary people on earth are the ones living an abnormal life. We are truly extraordinary beings.

I was born in 1959 with Saturn in Capricorn like the ones born in 1989-90, which means I have come ahead of the crowd, not only to keep open the windows of alignment, but also to maintain a consciousness free of fear. The physical part of living is the tip of the iceberg of my or our existence. I knew I was ahead of my time by a decade or two, waited for the subtle change ahead. My guide, Saturn in Capricorn, is timeless and ageless. Make no mistake! Saturn is the lead planet in many ways.

I am not challenging the beliefs of those who feel they still need to cling to them; that would be futile, since their minds do not comprehend this challenge. Challenging a person with beliefs is irrelevant when the house we live in is burning. Let them believe what they want. We are not here to question beliefs. We are here to dismantle them from within.

I keep my distance from the arena of beliefs. I feel they are radioactive and full of fall-out from the dead-zone. This is why some of the indigenous people told me that cities now smell of death and the modern society will soon die. For me this is like a deadly radiation type energy which destroys all of life. It is why I had to let everyone go in my life, except the shamanic few who were hard core enough to see the truth and are able to deal with it.

This is an operation for transition from one age to another, from belief to knowledge. It is a shamanic rite of passage for the entire world; in fact, the whole Cosmos is changing, and this is not something that is easy to deal with. It will take courage.

I do not know from day to day what is coming; but what I do know, based on what we have done so far, is that the future is going to be difficult. What I mean precisely is that it is *beyond belief*. The mind meeting this change is going to have to be utterly ruthless and as clear and sharp as a diamond. Forget recreational drugs, like

Ayahuasca, Jedi inner discipline alone will maybe cut it.

We must have clear sharp minds, or at least minds that can stay with *what is* and not with what they believe in, which is the denial of what is. They cannot be minds that are belief co-dependent. I was trained to use my mind beyond fear, to burn away every illusion of the modern world, to *see* with the eyes of a seer. I see where energy is wasted, and I assess potential damage, making changes in my approach to conserve energy. This is a time of direct action, not discussion. We cannot argue and discuss; we must know, and knowing we must act.

Humanity is up the creek without a paddle because real time astrophysics was cut out of our learning cycle when religions replaced the science of cosmos. Ancient cultures *knew* with certainty as in scientific knowledge, what I have been writing about for years. They knew we came from deep space, from the stars, that we are extra-terrestrial by definition, and that our destiny is to return to the Cosmos to move on home.

The vocabulary of the future will be: direct action; immediate use of your own intelligence; act now! Make Contact to Cosmos and survive! This is in essence what the astronaut Mitchell explained to me. We must re-invent our knowing and journey to the stars. For me, the first step of that journey begins within.

The final Karmic balance will come in the form of the special ones – the returning of those ancient beings who *know*. It is the colonizer, while killing the earth, who will face their own children looking out at them with loving wisdom... old souls in young bodies who do not believe in the future – they *know*. And so, in the days leading up to 2013, the usurpers of power will look back into the eyes of the infinite intelligence they once tried to destroy.

In 2004 a young boy from Russia made world headlines because of the unusual nature of his own private knowledge. Nothing that anyone had taught him, just information he knew for himself. His story first appeared through *English Pravda,* in which his mother said that he started speaking whole phrases at the age of eight months. The parents gave the baby a meccano, and the boy started making geometrically correct figures from it, combining different parts with precision. *"I had a very weird feeling that we were like aliens to him, with whom he was trying to establish a contact with,"* the boy's mother said.

"Boriska was born January 11 1996. He lives in Russia, north of the Volgograd region. The *Boriska Boy from Mars* talks of Lemuria, and his previous life on Mars, and of the coming earth changes. He also talks with great passion and enthusiasm about the Mayan civilization and how life will change once the Sphinx will be opened.

"Boriska told us about his previous life on Mars, about the fact that the planet was in fact inhabited, but as a result of the most powerful and destructive catastrophe had lost its atmosphere and that nowadays all its inhabitants have to live in underground cities. Back then, he used to fly to earth quite often for trade and other research purposes. It seems that Boriska piloted his spaceship himself. This was during the times of the Lemurian civilizations.

"Most interestingly, Boriska thinks that nowadays the time has finally come for the *special ones* to be born on earth. Planet's rebirth is approaching. New knowledge will be in great demand, a different mentality of earthlings.

"Something is going to happen on earth; that is why these kids are of importance. They will be able to help people. The Poles will shift. The first major catastrophe with one of the continents will happen in 2009. Next one will take place in 2013; it will be even more devastating.

"Do you know why the Lemurians died? I am also partially at blame. They did not wish to develop spiritually any more. They went astray from the predestined path thus destructing the overall wholeness of the planet. Love is the True Magic!" [*The Boriska Boy from Mars*, Pravda Ru, March 12, 2004]

↑

Tools of Empowerment

Flight of The Blue Heron

Light Sound of The Rainbows

Awakening Vortex Energy

Children of The Future

Tales of Space-Time Vectors

Gateless Gate

Zones of Passages

Transmigrating Emanations

Flight of The Blue Heron

"But why should I mourn at the untimely fate of my people? Tribe follows tribe, and nation follows nation, like the waves of the sea. It is the order of nature, and regret is useless. Your time of decay may be distant, but it surely will come. For even the white man, whose God talked with him as friend with friend, cannot be exempt from the common destiny. We may be brothers after all. We shall see."

~ Chief Seattle, in December 1853

The heron is seen in Native American language like the phoenix or the eagle of the Egyptians. He symbolizes the transformation of events and of peoples as well as the transition of times. I have been working with herons for many years in mysterious ways. When I was living in South Florida, the herons, along with the superb dolphins, used to gather around me, teaching me how to silently read their language.

Every day they would come for walks with me or sit next to me on the rocks, while they looked for little fish. They would even speak or make little sounds with their beaks. The dolphins were actually talking to each other, and when they met me on the pier before dawn, they would make their pleasant high-pitched sounds. It was quite a sight as they jumped around and splashed their tail fins into the mirror-flat silent sea, to make their presence known to me. Old fishermen told me they never saw animals behave like this before, but then I have always had a strange relationship with wild and free animals.

The dolphins have a sense of humor, and would sneak up on me when I thought I was alone. During one pleasant sunset I stopped to look across the inland waterway leading to the sea. My mind was far away, as the sky turned a deep blue. Suddenly a dolphin surfaced and sent out a spout of water, to my delight, to say hello. I don't know how the dolphin knew I was there; but they always seem to know and act.

On the day before I left the US, on a Sagittarius solar eclipse, neither dolphins nor herons were there. They had said goodbye on the previous day, and were themselves already moving South. They too knew the eclipse was the critical time marker. On the day I left, before dawn, only one old majestic blue heron sat at the pier and walked with me silently over a 500 yard stretch along the beach. When I sat down to say goodbye to the sea that had taught me its secrets for a decade he flew away. Hours later, I myself was airborne over the Atlantic. The eclipse was a time to say goodbye and a time for new beginnings.

It occurred to me that the flight path from South Florida to Central Europe was a very odd route; why not just fly straight across the ocean? I guess the aliens have a deal with the airlines, and pilots must follow set patterns of non-interference. I saw Washington DC and Montauk, Long Island from high above during the long night flight, as long shooting lights and the stars accompanied the plane due East over Iceland. It was as if the extra-terrestrials were accompanying the plane safely home, carrying a small group of wise humans who had withdrawn their shamanic presence from the Western continent.

Then, as the European morning lights dawned, the plane descended in the Old World, having left the new lands to their doomed inhabitants. Later the hurricanes would come. The Elders had told me that something dark is being activated in alien bases inside the Mexican Gulf and that many conflicting energies are about to be set against each other. They said too many opportunities to correct our path have been missed in the past hundred years. The palm trees in the Seminole territory talked to my shaman friend and said: "Get him out of here. This land will be faced with its karma soon." Soon, can mean many things to a tree. It proved be very soon in hindsight.

Then, we talked to an ancient oak tree near a Russian chapel and the tree said: "He is welcome here, but the mountains of his home are awaiting his return." I asked where *home* is. The tree talked to another tree far away in the high mountains of an alpine massive, and said: *Above the line of trees... the rock of the magi...*

I was shown how trees communicate with each other at long distances, just as planets communicate with each other, and how trees can commune with planets. If we trust the higher luminous beings of the guidance that are beyond the spirit hierarchy, then we can actually be plugged into Cosmic design in progress, of which we are an integral part. All of us are Cosmic architects of our own design. This is an ancient and powerful concept.

I was also told by the Elders that the Earth is shifting into a strategic position in the galaxy, and that is why we earthlings are now of interest to the many groups of extra-terrestrials who are also trying to dislodge the regressive aliens. So, yes... there is going to be some *collateral damage*, and that is what the governments around the world try to hide from us. Once a year, leading shamans and I hold small spiritual retreat leadership seminars where participants learn and apply the things I was shown. The flight of the heron signifies the journey upon which we are now about to embark. Learn to watch nature, and ask questions; then you will get the true answers, minus the *dis-information.*

Nostradamus has written many key predictions, which I quote throughout my work, as we approach the denouement of an endgame scenario. What awaits us beyond that point when we begin a new game with new rules and players? In his prophecies, the seer predicts four assassinations in 2010 and a total war thereafter, using germs and chemical ingredients, or the "ionic rope," and a global meltdown involving Islam and the Western world.

I was never so certain of what he meant and so I used my own skills. The figure eleven is mentioned. 2011, Uranus in Aries, squared by Pluto in Capricorn, sounds definitely like war, followed by eleven years of starvation; the British Isles wiped out by contamination; and the races interbreeding. 2022 sees Asia's rise from the ashes – the nation with a smarter missile shield: *"The Island survives the fire through fortified islands in the air."*

Maybe the French seer was also guided by star guides? We will never know. Could a fortified island in the air also be a *Twinkler* or a moving space station? Not even Hollywood could script it more vividly. I am merely the messenger of this script. I was told to write it down for you as it was given to me. My message and the vision of Nostradamus share common ground.

When a race lives in abstract constructs rather than embracing the visions embodied in the Cosmos, society deteriorates. This is what has happened to the people on the earth during the last fifteen thousand years. The deterioration of a race does not happen overnight, nor does the effect of psychological decay show itself until the race crosses a point of stubborn non-existence.

Look around. Is the galactic species of humanity living according to the highest truth, or is the human being losing sight of why it exists at all? The planet Earth is being destroyed by the high race of beings who were once known as *the guardians* – we were once stewards of the earth, its protectors.

Water is the life blood of the physical spirit, and yet the water on the earth is losing its vibration. How can the people of earth continue once the physical manifestation of spirit has vanished? Look at Mars. Would it surprise you to know that this planet was once similar to the earth? Its beauty was unparalleled in the universe, and yet now we see it dried, broken and depleted.

Our own history as a race is connected to the destruction of Mars and the loss of the planets physical lifeblood called H2O, or water. Why, then, would we continue to destroy the second planet of our origins? The first decay is the psychic, the next is the blood and the next leads to the destruction of the heart. Following that is the destruction of the human mind. Once the mind has lost its sense of true origins, the soul carrier can no longer function. After the soul connection is lost, the planet is destroyed, as was the case with Mars.

Rather than see this happen, forces concerned about the survival of the earth human are gathering for this time of the crossing through the world vortex. It has been decided that the Earth is too important to lose, so the human condition has been accelerated in such a way that either we will survive the crossing or be obliterated. If the forces of balance between light and dark leave the earth, the physical human will then cease and a new experiment will begin. A billion years are nothing to the *Twinkler* stars. They are but a flash of light in the eye of eternity.

The years ahead will follow the movement of water and everything related, which is the sacred geometric structure of physical sunlight. The decisions we now take will effect our own journey across the bridge of timeless visions. Who and what we are, *why* we are, matters – infinitely. *Who* we are alters the course of events within the reflected collective vision of the earth, and all of us are nothing less than diamonds reflected in the eye of Cosmos.

Have you noticed the moving objects that look like brilliant stars but are in reality shooting lights that blink and flash, or twinkle? The big blue *Twinkler*, T11, seen at times to the left of constellation Orion in Gemini signalled:

"We have sent to you one magi, an individual who in our culture is considered to be the bridge across time. It is your decision what you do with the vision and advice he brings into your world. He is with us."

170

Light-Sound of The Rainbows

"Tu es responsable pour toujours de ce que tu as apprivoisé."

The fox in Le Petit Prince
Antoine de St. Exupéry.

"You are forever responsible for what you have tamed."

Subtle changes already started to become obvious in nature during the winter of 2003, before the invasion into Iraq when the world's weather was showing signs of a greater ecological change. Behind that change was a mysterious *sound* which small numbers of people were aware of, heard or felt. This sound could be called a signal to the physical body of the earth. It permeated everything.

Although the trees, mountains and seas, plants, stones and rocks also heard this *light-sound*, it was not a reality to most people, only to the highly aware and spiritually sensitive. The sound was real, not imagined, and it seemed to signal through the earth itself, into the trees, plants and the rocks. But its origin was from Cosmos.

Nature was listening, being aware, paying attention. Mankind had been warned to take steps to change their ways and by the time this sound appeared the necessary changes had not been made. Governments were doing business as usual, and individuals were following the insane path laid out for them by the institutional sciences, technologies and corporations.

Since 2002, the force field around the planet Earth has been changing. The human psyche, or what I call the background Cosmic psyche that transmits life to the planet, is being filled with a new resonance. If you listen to the stillness within, this fact will become obvious to you. As we transit the times ahead, the mental behavior of the past is not going to transmit through the ether of this new psychic vibration, which is why people were warned to change their ways.

Free will exists as a divine creation. In fact, before humans were created, God created *free will*, the freedom to choose to live against nature or to work with the power of nature. In the last fifty years we have trashed the planet we live on. Free will also entails the reality of learning from our mistakes.

The Native Americans have a prophecy that says we will know the end of times is here when the rainbows in the skies become very rare and almost disappear. An old Native American Prophecy says, *"When the earth is ravaged and the animals are dying, a new tribe of people shall come unto the earth from many colors, classes, creeds, and who by their actions and deeds shall make the earth green again. They will be known as the warriors of the Rainbow."*

An old lady, from the Cree tribe, named *Eyes of Fire*, prophesied that one day, because of the white man's greed, a time would come when the fish would die in the streams, the birds would fall from the air, the waters would be blackened, and the trees would no longer be, mankind as we would know it would all but cease to exist.

This account of the Cree tribe from Lelanie Fuller Stone, was told to her by her Grandmother. She says there would come a time when the "keepers of the legend, stories, culture rituals, and myths, and all the Ancient

Tribal Customs" would be needed to restore us to health. They would be mankind's *key to survival*, these are the *Warriors of the Rainbow*.

The Cree tribe nations say that there will come a day of awakening when all people of all the tribes would form a *new world of justice*, peace, freedom and recognition of the Great Spirit. But the Warriors of the Rainbow would, at first, face a lot of ignorance and hatred from those who live and propagate the ways of the imbalance of nations.

At the time I could not imagine what changes would or could interfere with the appearance of rainbows...Then came the intense spraying of thick trails across the skies of Europe and North America. These heavy trails are not the contrails that appear due to commercial flights. The planes that spray them are unmarked. The long thick spreading trails are called *Chemtrails*.

The difference between Chemtrails and the contrails has been extensively analysed and documented on the Internet over the past two years. There is no doubt that the spraying of Chemtrails has caused either *dirty rainbows*, or an increasing lack of rainbows in the heavily sprayed areas. I have often watched the Chemtrails making ribbons in the skies and wondered what kind of pilot would go up there and spray this poison. The answer came to me recently when I realized they do not know the truth about what they are doing. They have been lied to in the same way the world was lied to about many other issues, such as Iraq. I still could not imagine what kind of a lie would get these pilots to willingly spray those death trails across the skies day and night, while keeping the whole enterprise a secret. Then I figured out the deceptive glue that holds the whole enterprise together.

Some claim the pilots have been told that their spraying is a solution to the problems of the ozone hole. The secret official story is that something drastic has to be done, or humanity will die within the next 20 to 25 years, maybe 50 years. Therefore they are spraying reflective particles, such as aluminium, to deflect the harmful rays back into space. Does that sound right to you?

I am sure there are probably rays the global elite want to reflect back into space, but those are the rays of the new harmonic, the rays of the vibration of the 5th world, or the 5th sun harmonic, the change in frequency that transforms the DNA, the blood and the human psychic vibration.

I can imagine the global elite being terrified of a major DNA change happening here on earth – a psychic change so dramatic, people will experience a direct spiritual awakening. These people will suddenly know they have been deceived, and will immediately act to stop the destructive flow of events coming out of that black hole in the human thought process.

Imagine an awakening so subtle that a soldier suddenly looks at his machine gun or tank and sees through the *psychological prison* of war and death, into the deadly illusion and beyond. Then he no longer can associate with this weapon and the activities involved in carrying it. He finds it impossible to kill another human being...

Ancient texts and Native American prophecies have spoken of a time when the sun speaks. A different light and sound will begin to emanate from or through the sun and as it reaches the earth, it will initiate changes. The new sound will change the vibration in and around the earth; it will activate the light codes of the DNA.

Imagine you are an alien entity who has fed off humans for many thousands of years, and this new golden fire that is to awaken within the human DNA will mean the psychic negativity you feed from will vanished into thin air. So you put together a program of spraying deadly substances into the atmosphere of the Earth to block the frequency or the spectrum of light that will activate this change inside the DNA. It is the change you fear most, and so you design a *Trojan horse* type operation to block it.

The new sound emanating from the sun could be likened to the voice of the creator. It is a last chance for the people of earth to wake up and listen, and change their ways. But those feeding from terror, destruction, fear, negativity and wars do not want us to hear this sound, the *light-sound*, which for them spells the last supper. So, they give humans a Trojan horse: *Operation Chemtrail*.

The secret extra-terrestrial-government alliance convinces the service-to-self elite that their world will die unless they develop some sort of *geo-engineering* for their own safety. The elite then convince the rest of the sheep that this is all a necessary part of survival of the human race, or it is imposed by secret decree. Thus, the Chemtrails, and the pilots who perceive themselves as heroes. The story is so good it is kept at a high level of secrecy. Since when will spraying aluminium particles and radioactive barium save the world from destruction?

The official reason for spraying those materials into the skies apparently makes a convincing story, but as the years pass it becomes clear that what they are spraying is killing millions of people, slowly over time. If you claim you are saving the whole world, the conscience can call it

175

collateral damage. After all, you are apparently saving the entire planet and the human race. If people realise the world is headed toward disaster, there will be loss of control and things will change.

Meanwhile, back at the ranch, the real reason for the spraying, and the toxic content of the Trojan horse, is the enslavement of humanity through the mechanical suppression of its Cosmic evolution. The reflective covering of particles has been designed to stop the penetration into the earth of a new color or light frequency. This is a spiritual war fought with secret occult technology.

Have you not noticed the growing absence of rainbows? Or are there just less rainbows? Or they are maybe dirty looking? Or they are very weak? The suppressed orgone and cloudbursting inventions of Wilhelm Reich were designed to naturally trigger needed rebalancing of the earth's weather systems. I live by simple laws, and anything that destroys rainbows is a bad sign in my world. So, I would say: The rainbows and nature are telling me the truth, and the governments are lying.

We are the children of the rainbow people of light, and we are here to live according to that light. The sound, the special light frequency coming from the sun, is altering the human DNA and the mysterious frequency of the background psyche, which is also the *psychic force field.* Is this special energy part of the light message of the Rainbow People?

The Native Americans know that when the time comes when rainbows are no longer seen in the skies, you know it is really the end. At that time you will know the earth changes are here.

Fortunately for us, something else is happening within and around our planet. The sound is also coming from within the earth itself, which is probably our spiritual back-up disk, just in case things get really bad. This means that directly within the psychic force field the change is happening, and it is being heard. As it is transforming the background ether, it cannot be blocked. The DNA is also changing. I hope we can all breathe without falling down later, when we realize that the object hiding inside the walls of the city is our enemy.

When you understand the strategy of earth security you must also be totally free of fear. If you are afraid you can be controlled. If you can be controlled you can be driven onto the rocks, where hidden forces plan to control your future and the future of your children. Rather than being manipulated by these forces, it is time to waken up and take control of your own life, by acting as a sovereign individual, a sovereign being, rather than as a slave.

As mankind was born from the *psychic divine*, it cannot be controlled. However, only when man finds true compassion will he be master of his own destiny. This then is the challenge facing us in the months and years ahead. Find Compassion! Act upon it!

"Strange times are these in which we live when old and young are taught in falsehoods school. And the one man that dares to tell the truth is called at once a lunatic and fool." – Plato

Awakening Vortex Energy

"Any smoothly functioning technology will have the appearance of magic."

~ Arthur C. Clarke

The human description of reality is a culturally defined agreement on the meaning of images and symbols seen, interpreted, and understood as the keys to and icons of the culture we live in. In other words, in order to survive in the post endgame times, we will agree to decode the code in a self-empowering way. Reality is redefined via a deeper understanding of symbols, or glyphs. Thus, *knowing* is another internal or psychic awareness we use to re-define our world view. Indeed, it is the way agreements are arrived at in matters of magical faith. It is the *Magician's Way*.

Mystical experience is the active unknown. In his book, "The Way Of The Explorer", Dr. Edgar Mitchell wrote, *"The mystical cannot be experienced intellectually any more than one can learn to swim on dry land."* In other words, you cannot comprehend star magic unless you live it.

Mind and matter are not – as was thought and taught for millennia – separate realms; they are two inseparable aspects of one evolving reality, two sides of one coin. So too, are the star signs, their planets, the universe and humanity one single evolving reality.

Aleister Crowley *said, "The meaningless and abstract, when understood, has more meaning than the intelligible and concrete"* Science and mysticism were once long ago united in working with the void of the magical realm. As a magician I can see the two forces returning to the central stage of science, over the next few decades. Indeed, they need to come together, applying mind over matter, to discover the *power of the abstract*.

Matter *is* the mind. To focus solely on physical reality, has always been considered by mystics as an unstable and unreliable way to navigate through the landscape of life. Mind and matter, together-as-one, point to one single, ineffable, uncertain, and ubiquitously interconnected *super-reality*. A reality with two Janus faces: the two Gemini twin gods of physicality and mentality, or existence and knowing – dyadic pairs.

As Neptune and Chiron are making their way towards the Pisces sign by 2012 – we will see science and mysticism make peace with star magic; just in time for a *Nobel Prize in Zen-Magic*.

Egghead scientists and theologians will have no choice but to respect and recognize the magical 6th sense for what it is, the all important and essential navigational element, the most primitive, the most ancient and yet also the most precise. Obviously emanating from a sub-atomic molecular level, the psychic awareness delivers access to and from the deeper mind, the intuitive process, astrological energy, psychic recall, non-local resonance perception such as extra sensory perception, also the internal evaluation of mental and emotional well-being.

The psychic is tantamount to an advanced spiritual radar and guidance system. Without it, humans would not get far in life. Without it we are lost. I could not have made such accurate predictions as the 2000 ballot recount with Bush Presidency from a vantage point of Palm Beach in December 1999 when neither Bush nor Gore were even candidates. I could not have written this book without my cosmic feedback.

Magic is the art of producing the desired effect. Star magic is the exercise of this art together with cosmic forces, or supernatural agencies, such as stars, planets, and galaxies. However, I see nothing supernatural in harnessing the non-local resonance energy that can be made subservient to man via inner focus. Star magic is intention and awareness merged into one. Awareness is the perception of energy, while intention is the willful propagation of that energy. Perception and applied focus of energy equals star magic.

We are now all experiencing a time warp, which is causing rapid and multiple changes. Welcome to the magic circle of those now entering the awakening vortex energy. This change is neither good nor bad. It simply is what it is, and we would do well by welcoming the change, adapting to it intelligently, listening to the stillness within and by observing the space between the spaces.

Enlightenment or peace of mind is a rare, and yet you have to find that space to understand what is happening. The majority of human beings are being purposefully stressed out so as to lose their inner navigational compass, their sensitive psychic awareness.

It will necessitate the cumulative perspectives of large numbers of the wisest people to explore the potential facing us over the next decade. Humanity must wake up now. By the time most people grasp the significance of what must be done, it may be too late. This shift will be a complete departure from our previously acquired thinking.

We are capable of knowing and applying strategic astrology today, with Uranus in Pisces, awakening the masses in shocking ways and with Pluto in Sagittarius, transforming faith radically. With Neptune in Aquarius, the solutions work globally if not universally. Unlike the Mayans, Vedic experts and others, whom I deeply respect, I am not interested in the known inner seven naked eye planets, because my world begins where theirs ended long ago. I have a feeling the radiant Atlantean magicians worked with more planets than we know of today.

The great traditions inherited from ancient thinkers and philosophers are anchored solidly as a starting point, for us to apply to our individual lives, and to a Cosmos we can re-invent and re-design as we go. This requires two things: knowing and imagination. We are obviously not going to find either of these two valuable skills in the social system contract we have entered by being born at this time. But we can develop them nonetheless.

What contractual system do I speak of? To date, this intellectual exercise called spirituality and existence has proven too challenging for most of humanity, who seem only capable of satisfying materialistic desires. The rest of us are on the brink, perhaps out of necessity, of lifting our eyes to the star gates and beholding any one of the planetary beacons provided by the creative forces of Cosmos; giving us a global frame of reference for our natural existence. In other words, soon we have to shift dimensions.

When most of us were born into this earth system, during the 20th Century, we were handed – without knowing or necessarily agreeing to it – a social universal system contract of human behavior. This agreement was worked out by the powers who seek to be in control of the planet. Alien races are also involved, one of them being a regressive reptilian race who lacked evolving gene pools. The system of this contract goes back many centuries, even thousands of years, with the emergence of dark priests intent on enslaving mankind to their will.

The system's agreement to which most people adhere is commonly referred to as the *matrix*, or of the global big brother elite organizations. One example is the NSA, which is controlled by reverse engineered alien technology. These regressive control systems only exist if we allow it. Most people on this planet adhere to this deal. At present we are the blind following the blind.

The powers that handed you the agreement never told you so, but they don't really intend to keep their end of the bargain which is *your security* – not that the deal itself made any sense to begin with. But even the deal that was promised to you will not be delivered. This is why I left the system long ago; I foresaw the breakdown of the earth as an outcome of the disorder we create *within the system*.

The *deal* and the *agreement* is the structural architecture of this society we live in, the planet we inhabit, and exploit. The ancients would say this earth is a place run by fallen angels, dark forces and entities. Darker forces who believe they can scare humans into accepting loss of freedom for greater security.

Too bad humans don't realize that they are more powerful than the fallen ones. I would simply say we are living, at present, in an insane asylum called earth, with a few insane people in power. They are running an asylum of six billion willing people who agreed to be deceived, abused, controlled and manipulated.

I would also submit for consideration that the elite human controllers are in the grip of a dark force with which they also made a *deal* of sorts. Most of the human population have joined their way of thinking by not paying attention to their own awakening. This neglect is a one-way road leading to a dead end. As long as we agree to this *social contract*, we contribute to the insanity.

As J. Krishnamurti so eloquently said: *"It is no measure of health to be well adjusted to a profoundly sick society"* At birth we enter the deal of society, and have to decide whether we live according to its insane ways... and in doing so *betray the earth*, or do we live intelligently? From the moment we accept the contract of betrayal, we pretend to have a good education, which is system sponsored, produced and monitored, both public and private. System perpetuating *knowledge* – the system's operating programme and agenda – will be delivered to us, together with the belief system called *religion*, if we take that road.

A constant stream of audio and video hallucinations and what we are told is *enjoyable* music, matrix type movies and most *educational* television shows condition the mind to sleepwalk through life. Through this rapid eye movement sleep deprivation media, we are subtly programmed if not mind-controlled to be good system participants. The system's *spirituality* is fear based. Meanwhile people are fed genetically modified organisms, to further destabilize their minds to work against nature.

The advertising industry makes you believe you should drive a fancy car with an impressive sound system – or you will let the car drive itself with a built-in auto-navigation, so Big Brother can find you better, in case you get lost. Never mind the fact that soon there will be no streets to drive on, care of massive earthquakes, landslides and flooding.

Your children will be dumbed-down via hi-tech vaccine scheduled subconscious implants. They too will raise equally dumbed-down subscribers to society. Your security, and that of your family, extends seamlessly from the cradle to the grave, and then beyond to theirs. While the awakened ones shift to a non-matrix existence, like the elves before them.

The alternative that lies before you is: Pay attention to the *Special Ones*. Some light beings will appear physically in this world. Most will work from the invisible and inter-dimensional realms. It is not beyond you to see, hear and talk with the invisible realms. Guardians and co-workers of the human race do exist, but *you* have to develop the sensitivity to work with them.

The eye in the sky may wish to watch over all of us *mercifully and benevolently*. Its central switching computer can transmit an electronic script of obedience to a specified subject at a pre-specified time. The script's transmission range can be limited to a single building or city. Perhaps later extend to cover a large geographical area like North America, Europe, or both; or perhaps the entire planet.

In exchange for behaving well and for being taken care of by the system, you are to surrender your freedom. This means, no freedom of speech, nor of the press. No right to assemble, nor criticize the state. You will not be secure in your person or house from searches by the state. You will be permitted to travel only when and where the state authorizes you to.

Heading towards the worst possible future you will have no right to start or run your own business, nor the right to be your own boss. You will be forbidden to invent, think, write, draw, paint, sculpt or create in ways the state deems disruptive, and you will be forbidden to bear more or less children than the state has authorized.

Shortly after the whole world is governed by this system, the world economy will collapse, just as Rome and the Roman Empire collapsed. The cities will be wasted as the Mayan, Shamanic and other predicted signs in the heavens will have occurred, and almost everyone, including nearly all those who entered into this world system will die of disease, famine, or violence. Few of us will survive.

Chaos will prevail for a while until things normalize. The system's behavioral modification procedures will have been useless. This is why I advocate you psychically leave this deceptive system now. Beginning in 2009, we can expect the effects of the system's mismanagement to become evident throughout the world. It will be like a mental disease, starting inside the US, and it will grow like a cancer, outward, to engulf other areas.

The arrival of Pluto in 2008 in Capricorn will probably be the harbinger of *Contact* with extra-terrestrials. There have already been clear signs that non-human races are watching us. Contact will also include the acceptance and practice of higher level astrology. There is not much time left to make a decision whether to stay inside the *matrix* or opt out to create your own intelligent solutions.

You must comprehend that what you perceive as reality is an inversion of your natural self-navigation senses and knowing. Let me explain this in a more subtle way, as I show you a doorway through the inner psychic reality. What we perceive about ourselves is a mere construct of what was once real and truly powerful.

Genetic manipulation has changed the reality of what we perceive, into a projection. Via mental imagery we think we perceive what is real, when it is only a fragment of reality. The bigger picture includes an awareness of the invisible realms, the spiritual realms, trans-dimensional realms... need I continue?

With extremely limited awareness we are trying to make sense of what we can actually not understand. We try to cope with it, but we do not actually comprehend who we actually are, and the system uses our ignorance to its advantage.

Reality has been switched on humanity, mirrored, inverted, turned inside out and upside down. We live as if inside the negative of a photographic film. Our existence is like a carbon copy, but it is no longer the original it was meant to be.

We actually live in an illusion. We can recapture our original power by first realizing the illusion created by our senses and our faulty mind capacities, and then by beginning to reengineer ourselves from within. No one will teach us this process; we must learn how to do it – self navigate – for ourselves.

You are your own power. You must clearly understand this. No one can do for you what you can only, and will do from within. The colonial system is here to trick you and to make of you a slave – with your cooperation. As a result of realising this Cosmic law of the inversion process, people tend to look outside to the system for answers and guidance, when *we are the answer*.

We are supposed to journey forward into the channel of energy, not go backward into vertical time. We are the energy we seek. Once you see this, the inversion is defeated. Peace will then follow naturally.

"Observance of Source in All Things is the principle that all manifestations of life convey an expression of First Source. It does not matter how far the unifying energy has been distorted or perverted; the Source can be observed. It is the action of perceiving the unification of energy even when the outward manifestations appear random, distorted, unrelated, or chaotic." [Wingmakers Philosophy: Chamber One. *"Observance of Source in All Things."* From, Wingmakers.com]

Children of The Future

"Beyond a critical point within a finite space, freedom diminishes as numbers increase. The human question is not how many can possibly survive within the system, but what kind of existence is possible for those who do survive."

~ Frank Herbert, Dune

I have not as yet commented on the extra-terrestrial, or shall we say the *invisible* world side of the Cosmic conflict, from the breaking of the original knowledge into its fragmented parts. As an astrophysicist, I am well aware that the philosophy of astrology is maybe older than time itself, if we care to look at it in proper or ancient ways. Astrology is the mother of all sciences and it is the *cosmic language* per se. It may predate earth itself. To me it is the visible manifestation of the original frequencies.

We will soon see the day when this cosmic reality is the leading science again, minus religion. Ancient savants and sages were taught astrology. The celebrated Jesuit thinker, Pierre Teilhard de Chardin expounds on this in his theory of *cosmogenesis*. I am not an expert at everything, but via astrology I can better understand what relevant questions to ask, and so I have become quickly knowledgeable in many areas. I have used this open-ended attitude to collect information and write my books.

In some way, astrophysics and the passage of Uranus (ruling astrology as a science) in Pisces, the sign of compassion and cosmic mysticism, until 2011 will certainly shed light on the following issues. Let me address some of them. I want to look at many subjects as well as different areas of our globe and of cosmos, as I see them in relationship with one another during these planetary movements. I would also like to present many viewpoints, in order to provide an adequate or appropriate picture of the major puzzle pieces.

The dominant religions on earth are really all from one school of thought. The original scriptures were hidden as fragments of the original school began to separate and vie for power. Each separate fragment *thought* they were right and the others were wrong. However, they all came from the same source, and in fragmenting the knowledge they perverted the original teachings and destroyed even their own small fragment of *truth*.

We must find our way back to the original source, the source material, and then step beyond it. What I love about my time portal doorways of communication between the worlds is that they cannot be destroyed, fragmented or changed in any way. Theirs is not a written language; it is a living language of truth and light. The wisdom on the earth fragmented and was lost, but not in that other world. There it remained, as it is now, complete.

The human soul is the object of the war of fragments, and humans themselves are now the battlefield. Where did we come from, and where will we go? These are age-old questions, whose answers are written in the stars, near the 13th sign constellation of Ophiuchus, the serpent charmer or Christed One; near the Southern Cross.

Today's world is obsessed with individual saviors, or future messiahs and the peace they will bring to mankind. The fact is, without peace now – as in today – there will be no future peace. It was tried already, a very long time ago. Light beings came to earth to *save* their fallen friends and comrades. Perhaps this is where the imagined story comes from. Their attempt failed.

Why did it fail? Because, their fallen earth-bound friends did not recognize them and turned against them. Then a new battle began – the earth-bound souls against the light beings. These *rescuers* had to face the bitter truth that the ones they loved had become the enemy – in some way that's you and I.

Those who had fallen to earth had taken on the sickness of the human race they had come to *save*. Then the next wave of saviors arrived to rescue the first earth rescue group. The earth-bound ones saw the following wave of *savior-rescuers* as intruders. The confusion was complete.

The ancient wise ones then concocted a plan to fit the strange nature of this new problem. They formed a highly complex network of other dimensional, extra-terrestrial, and spiritual beings whose aim was to teach, awaken and support the next wave of rescuers.

The changes needed – it was realized – could only be triggered from within. Of course, the dark force does not want this and so a strange battle is underway within the human beings themselves. As I said earlier: *humans are the battlefield*.

The special ones I have talked about are the ones in the future who will outplay us all in the chess game of life. They are the ones who will refuse to live a life of violence. By that I mean violent thoughts, hatred and dividing the inner from the outer. They do not need a messiah to come and save them, they are the rescuers caring for and saving the earth. They will only feel sorry for the people who put aside reality and their planetary responsibility in order to be part of an insane world.

These beings know how to live, and will refuse all programming and indoctrination of religion and the fragmented hierarchy. For them the time they have on earth is about action, doing, and being who they are meant to be. They know no compromise. They only know love and their love is terrible, all encompassing, free of the shadow of doubt.

These beings are not held back by dogma and belief. You don't need to believe that which you clearly see is true. They are the future of the earth. They treat the last dying remnants of the religious orders with a silent awareness of a dwindling mass who appear to be mentally deficient and physically incapable of action.

Cleaning up the mess? For those who are self-responsible participants in this world – the new age pronouncement that is echoing around the halls of the inner brain cavities about the coming *children* who will lead us into a new world? I call it escapism.

I don't know if one can find a diplomatic way to say this? But, escapism is claiming that the *Christ*, or *Indigo* children are our saviors. Just stop and think about this! It gets interesting as we give away our power to an idea.

A beautiful kiddo, kind of like I was, comes onto this earth; maybe many of them enter the physical reality all over the earth. Are they emissaries and Ambassadors from Sirius? Pleiades? Aldebaran? Andromeda? Or are they from our own Humanity? The kiddo, and many of these children are wise, peaceful, grounded and in touch with Cosmos.

Now bear with me! This brutal warlike human being, who has depended psychologically on organized religion and political leaders, but who has not worked to transform the inner confusion, the anger, the division, the fear and aggression, announces to the world that beautiful and advanced beings (children) are here on earth to clean up our mess? Or lead us into a new world?

So, you have this beautiful child, and you use them to clean up the political, psychological and social divisions you have created, to clean up your inner psychological division. How? If it was possible, don't you think it would have happened already, by now? The rule is: If you create the mess then you clean it up, or drown in the sewage of your own creation.

We are here to signal to those who wish to do the work to activate an inner not an outer time-space alchemical change. What I mean is that the human being (the self) has lived, until now, a life of confusion. The inner conflicts are unresolved and the search to find an outer solution to the inner problems has led to nothing. In the meantime the *confused* have utterly destroyed this world.

The psychological (karmic) pollution is increasing because of the growing unresolved inner conflicts and fears. So what do I do as a responsible entity living on this earth? Do I announce that soon the savior is arriving, or special children whose task it is to help you and me to evolve, to help you to resolve your destructive way of life, or to help you to change? Free will means the choice is yours and yours alone. It has nothing to do with anyone or anything outside of your own freedom to decide which direction you are moving in.

Who says that these beautiful children are here to clean up after the mistakes of the old brain? Who says that they are here to occupy their minds with problems of the past? Who says that the ones arriving have anything to do with us? Maybe they are here for another reason. Maybe the age of the problem makers has passed, and something new is about to dawn.

It seems that the problem makers are obsessed only with making problems – service to self. The *new age*, to the problem makers, means that anything new that arrives on this planet is here to occupy itself with their problems. The idea that any new intelligence is here on this earth to solve the problem maker's troubles is short of idiotic. Perhaps these beings are here for another reason that has nothing whatsoever to do with helping the old brain pass into the new world.

The way I see it is that each human being, or individual as a sovereign being, is responsible for his or her *self* and for the transformation of *self*. I do not see that any other being is responsible for the transformation of others. A teacher is someone who points the way to the inner transformation. A teacher is not someone who is in any way interested in transforming or alleviating another human being's problems, phobias, fears or conflicts. If the inner bridge is not there then I doubt any child can build or repair it for another. That is why I say: *Leave the kids alone!* Let them be! The new is the unknown.

I see a future where the problem makers bring all their problems to the children or to me, and they are politely asked to: *"Please leave us alone, because we have work to do."* This does not mean that we do not work together, co-operate together and help each other. It means that the way we work together is different than the dependency based system in place today.

I do not see that the *special ones* are in any way related to anything the present occupants of earth are doing, creating, building or holding onto. How can a being who is outside of time communicate with a being attached to the illusion of psychological time? The two cannot communicate. Therefore their current manifestation on earth has to have another function, another reason. But those caught in time cannot see this, they can only think of themselves. The self being the product of thought, and the soul is something outside of the world of thought.

Most people, not all, but most... live on planet Earth as though they are squatters in a house while the owner is away. Not caring how they live, they fill the house with garbage, toxic waste and their accumulated filth. They dirty the water and leave the house to fall into disrepair. When some bright child, who cannot stand the toxic stench of rotting decay asks the other inhabitants of the house about the mess? The other inhabitants tell the child that: "One day the owner will return, clean up the mess and restore order."

If you imagine that the house is the human mind and physical body, and you *are* the owner, the solution immediately presents itself. Those *returning owners* are the beings now arriving on earth, and they do not intend to wait for someone else to create order. They intend to do it for themselves.

Tales of Space-Time Vectors

"By 2010 – Humanity will be externalizing the core of the inner heavens."

~ St.Clair

During my earlier childhood I experienced at times, when I was less encumbered by too much *knowledge* from adults, the totality of my own life as one single master-moment, past lives included. In astrology, this means the complete hologram of stars and planets at birth in progression and motion, as one concentrated force, which is what it is... one vector into space time. I saw my whole existence again when I met *The Master of Light.*

It is so for every human being, but only few can feel, let alone understand and decipher the spiritual hologram of this core energy. Time is an illusion. This does not mean that our tracks lie ahead of us as set realities which cannot be altered or changed. The hologram is not a finite set physical reality. It is a dynamic energy woven within us as colorful as a rainbow and as varied as the whole universe of stars.

The physical time of birth is the holographic imprint of Cosmos. When the human being draws his or her birth breath, one looks to the East to see what sign degree rises. In my case this was 23 Sagittarius, which carries this Sabian symbol: *"A Group Of Immigrants As They Fulfill The Requirements Of Entrance Into The New Country."* Maybe this location, earth, was the place I was supposed to enter at that time, coming in with the star knowledge of the extra-terrestrials. When those of us who remember both past and future come into the earth cycle, we also know the map of the world and how it develops.

So-called powerful nations, their leaders, and a series of countries will collapse. So, of course, will their *affluent* companies. The markets will crash, and businesses will crumble with the house of cards of the non-existing economy, for no apparent reason. Yet the reason will be a cosmic and a collective *core energy movement* helped by a few weather changes.

When I was growing up in the land-locked country called Switzerland, I knew I would wind up living by the sea, on the shores of a beach. That was the image I both received and sent out. The mirror, in-out... and of course, I did just that, I lived by the beach. Now I live by the lake. As a little kid, I used to say to my parents that I wanted to live in the big tower by the lake, and that was the first and only apartment offered to me on my return to these shores. Now it is my magical Rivendell tower, overlooking the *Mirror of Galadriel*, my lake.

Some prefer the misty mountains. Either way, our new lives will require silence and peace. We will have to find safe spaces outside the matrix. Maybe we shall create the *City of Gold* on a mountain by the sea, if not directly at or on the sea, then looking at a sea of stars. Perhaps it already exists in times ahead of us. Water world: Uranus in Pisces until 2011 and then Neptune in Pisces until 2025, now the prevailing and mystical signature for a long time ahead, during many years, which will mirror down to earth several unexpected changes of the sea and landscape.

Chief Seattle said that Earth does not belong to man, but man belongs to Earth, and whatever we do to the web of life, since we are a strand and part of it, we do to ourselves. Earth, another core energy shell, is therefore our host and our home. Now we have to make sure that we have an earth left worth living on past 2007, when I feel things will become a bit unusual, to put it mildly.

The next decisive years are about *anticipatory guidance*, or how we manage the core of our holographic shell, the inner abode, and how we nurture it... to connect to the Uranus spiritual web of life in Pisces. This is what we are preparing for – the magi's passage. I have learned an oddity: No one can tell me who I am in the future. It is a living dynamic that I am creating now. In fact, I have been shown there is no future and no past; just an ever present now that contains all realities in this moment. They are contained as potential in a birth chart in motion.

On a number of occasions, who I am to be has come through some kind of time portal and has contacted me. There is a feeling of genetic mutation, or evolution happening inside the magi's illuminated passage that the people of the future have been waiting for. What is given to our birth remains hidden from us until we actively ask our guidance to activate it.

We are a kind of living dynamic, and I know some part of me wants to create that other reality here and now, which is a totally different life than the one we have at present. Yet, my wiser levels of being tell me to get on with the current passage. What I am shown is an emergence that will take us all into a new future. I have been through the same thing before. Indeed I have seen and done this work many times. It is *Merlin's* or *Gandalf's* battle against *Mordor* type of reality.

The wiser persons of history truly never had to say as much as one word, and people would soon find out for themselves that they could not stay within such an intense energy field. As we are relocated by the spirit, wealth will be valued in terms of human happiness and not in terms of *service to self.* The guidance of our ancestors or – *who we are across time and space,* are bringing us closer to this unusual energy field. The wiser magi – the one who looks back at me when I gaze out into the middle distance reality, has an incredibly powerful, yet patient energy around him. He is somewhat peaceful, albeit persistent, nudging me along.

Times will become weird. The net of *wyrddyn* (Merlin) is a complex of abstract opposites. The contrast of that peaceful silence and immense force is what I was told is called a *morphogenetic evolution.* The light force is activating inside those of us who will make the genetic evolution happen.

There is no time to waste if we want to see the *City of Gold* or any other *Islands of Light* manifest. We are going to go through some tough times emotionally, and some of us will often feel we are being "burned away" through these feelings. We will learn to navigate through life in new ways in order to create and transform.

The *City of Gold* is not for the faint-hearted. The spirit mirror showed me the reality of our sad world, and the consequences effect all of us. You don't get to "revelation" type levels cruising through dependence on others who you hope will create the order for you, despite what some might believe.

As we ready ourselves for the awareness shift or the upgrade in consciousness – *The Magi's Passage* – beyond the realms of judgment and blame, to *know* the oneness of being, we have no room for error. There are no mistakes allowed, nor are they possible. There is nothing to judge or even feel bad about. We do not judge; we *see*. The self-imploding games of thought are transitory illusions that hold us to the will of self, and keep us from experiencing the oneness of being – a concept astronaut Dr. Ed Mitchell deemed of vital importance when I interviewed him. He also suggested we have to actively create our future, not sit and *wait* for it to happen.

I expect more unexpected events in the outside world, as certain human-made prophecies are played out, against better judgment. What is about to come into our life will sweep into our soul and our deepest feelings like a great river. I was advised – and I pass it on to you – to concentrate on the Piscean vastness, and not to get hung up on the small Gemini surface things. Now the Pisces Uranus must find its way through the Cosmos in sweeping manners connecting to you until 2010. Then you will have to swim through Pisces with the help of Neptune until 2025. You do have one planet in Pisces at all times for twenty years. Use that time well.

The Master of Light, who communicates with me inside the mirror of the soul, is a living dynamic of the integrated experience of himself across time and space. He knows and therefore he is. He is my point man or teacher of awareness. Why is the vortex so strong around the beings we are in the future? Because, *who we are*, allows the force to move through and around us freely without restriction. This is also an important part of the solar wind energy of my natal Mars in Gemini at birth, so I am told...

What is the *inner guidance?* Within everything is the seed of everything. On the other hand, from nothing – nothing comes. *The Master of Light* showed me that in each small particle of cosmic dust are concealed all the elements of Cosmos, and thus of man. Since these star and space dust fragments are trillions of times smaller than an ion or electron, these seeds will have to wait – as everything else that is unrecognizable or incomprehensible – for the time assigned to them for expression and growth.

The time lord and magi know that growth can occur only in two ways: either by nature or by art, naturally or artificially. Bearing this in mind, we inevitably approach the age and art of cloning humans with all its possible consequences. By asking for guidance we learn to do things the way nature does them, not by imitating nature, but by placing ourselves within nature.

Since particles are the building blocks of everything in the universe, there is a necessary link between the behavior of these minuscule units and the objects composed of them, including our minds and ourselves. In a curious twist of the inexplicable link between all things, experimenters who are searching within the brain's structure for a physical foundation for the concepts of the thinking mind and human consciousness have discovered that the brain contains structures called micro-tubules.

These substances contain an electron held in the same indeterminate state as the particles that make up everything else in the universe. Furthermore, the position of this electron within the micro-tubule cavity determines the protein composition of the molecule and the function of the micro-tubule.

Here is the curious fact: If the electron is paralyzed within the micro-tubule, as when a gaseous anesthetic is used on a human being, that person loses consciousness. From this bit of data, experimenters have extrapolated the theory that it is the moving electron in the micro-tubule that gives rise to consciousness and the thinking mind.

At this point, we enter unknown territory. If the conscious mind indeed is the result of a quantum process, then the indeterminate electron in the micro-tubule or the cart in the corridor, and our whole perception of reality is a function of mind. Then that perception is ultimately grounded in the quantum process with all its quirky behavior.

How does this information effect our experience of events around us? When considering the past, for example, could it be that what we perceive as the present is brought about by an act of mind functioning through a quantum process and creating experience by "collapsing" the many possible options before us into one determinate event?

In that case, what comprises the future? Could our ideas of the future merely be our intimation of the possible states of experience held in their potential form within our own minds? Could it actually be, as we are told, that if we visualize success with some endeavor, we are actually creating the reality we wish to see?

Something ultimately happens out of the transformative ongoing soul mutations, as the future remains an amazing enterprise. This is why I say the *future begins now...* I see it with crystal clarity when its presence comes to guide me. The same power inside us now is at the full command of who we are in the future.

There has to be a high level of order in the people around us who will manifest this future *City of Gold*. Service-to-self individuals will not be able to live inside the energy of peace and the oneness of heart and mind. This intense force will not tolerate corruption. Integrity burns its way into the soul in shamanic or magical ways. It demands everything, and leaves no stone unturned.

The way *The Master of Light* sees the space-time vector is what we know from our own future yet to be. Before we take our first breath we usually have a memory of what went before. We reactivate this memory by asking our guidance to show it to us. When that happens we adjust our navigational skills as to what choices we will make in the now. This can happen at any moment once we integrate guidance into our being and seeing.

Gateless Gate

"Humans are amphibians – half spirit and half animal. As spirits they belong to the eternal world, but as animals they inhabit time."

~ C. S. Lewis

The teachings of the monoliths and star gates on our planet are very sketchy at best. At worst, the knowing of the gateless star gates has been hidden so we are stuck here, for now. This "being stuck" is also a self-created illusion, arising out of our non-human behavior, and with the change of realms we defragment the fragmented hard drive and correct the error.

I would imagine they are building a wall around the Pyramid of Giza, because things are about to happen in many weird and wonderful ways. But as the weird and wonderful seems to frighten this modern man, it has to be hidden behind a wall.

We are the Atlantean and Celtic lineage of Egypt, the same line as the many magi healers. All of them, both men and woman, were spirit light incarnations of the same force more commonly known as *love*. Albert Camus wrote that, *"Man is the only creature that refuses to be what he is."* The truth was hidden to confuse people. They were all magi – the people of the light. We are the children of the mystery, and the mystery is waiting for us to awaken from this little nightmare and reconnect to who we really are.

Babylon originally meant, *baby lion* and refers to the Syrians or the extra-terrestrial *Sirians* who are part feline, part human, and who gave birth to a cultural heritage long ago buried under the vast sands of the Middle East. As well documented in the books by Zecharia Sitchin, the tablets of Sumer are the original source of the Hebrew and Christian religions. This is the material that was later used to build the *Re-illusion* of our present time, and create its own *false truth*.

One could say that we magi came from the cat people. The *Crystal Light*, in interplanetary language, is a star gate or pyramid, and the sphinx is the Goddess Bast, as suggested in *The Pyramids of Montauk*, by Peter Moon and Preston Nichols. They suggest that originally the Sphinx had breasts, and the face was female. Crystals are needed for time travel.

The nature of the *star lion* is a harmonic sound and one cannot pass through the star gates without it. It is a natural part of us, encoded within the DNA. It is called the *Lion Gate*; and both the Scots and Iran have a lion on their flag. That is why two lions stand on either side of pillars, or Gates. Ishtar, named after fixed star Spica, is the cat people's central refuge.

A sound once emanated from the Egyptian sphinx, and from the pyramids. Only after Atlantis, did knowledge break into the different streams we know as the mystery schools. Each fragment formed its own *secret society* and each took an aspect of the truth and fragmented it further. It was then distorted to create leverage for emotional and mental control of the masses.

The child like fear of *the beast* is fear of man's own shadow. Those who govern wanted to split the world into higher man and lower beast, separating man from nature, and from his *own nature*. They also wanted to destroy the race or lineage of those who were descended from the Syrian feline race that established their home on Earth after they had left Mars. We will one day discover that there are similar architectures on Mars and Sirius.

The architects are the *lion race*, the line of the Pharaohs, Akhenaton, Moses, Christ, the Cathars, and the Celts. These are the bloodlines that can easily pass through the lion gate, or the star gate. They are the navigators of the heart frequencies. C. S. Lewis refers to these things in oblique ways. I like to spell it out.

Throughout my life, and this also happened often in my childhood, I would awaken into an altered state of consciousness, into another reality that exists simultaneously with this one. Thus, you see the room you are in, but the room is not made of solid matter. You perceive it as a *vibration*. To move out into this other reality and navigate it requires an almost animal-like awareness. Thought, the *small ruler* measuring time, is terrified of this other state. Thus, the rational mind was easily programmed by religious doctrine to associate higher awareness with a non-human *beast*.

Sirian star beings navigate through this vibrational doorway or portal because we speak its language, the language of sound, which I now understand, after all these years. When I am in that world I open the *doorway* with a sound. The sound comes out of an area around the solar plexus, and is similar to the roar of a lion.

The inter-dimensional door opens *only* for that sound, which is a precise feat of spiritual engineering. Only then can you pass through and leave this world. The harmonic resonance or frequency protection is why the other alien races cannot access or use the star gates. They do not pass through because they cannot emit the trigger frequency to open the frequency of the energy portals. Those with Sirian DNA are able to move through unseen timelines.

It is possible that we are able to travel backward and forward between the original time line of the *Creator* and this one, thus making us children of the *Christ*, not one Christ, but the actual inter-dimensional race of beings who use that name. We would then be teachers and holy persons capable of showing the human race how to re-turn their reality back to the original time line.

Those who wish to maintain control do not want humans to awaken to their true origin. The deception programmed into the education system about the B-East is to make the ordinary person afraid of those who can move between the time lines, as did the Sumerians, Syrians, Assyrians and others related to Mars, including the Cathars and the Celts going back to Atlantis and their Atlantis Oracle.

The other instinctive side of us, the ancient side, does not speak a linear fragmented language. The psyche or the *psychic* uses whole complete sounds. The roar of a lion contains multi-holographic levels of sound, most of which we do not hear. Did you read the books by C.S. Lewis, about *Prince Caspian*? Just think about it: the Caspian Sea, Aslan, the creator lion, and king of the beasts. There is much more, but I think the message is clear.

On the other side of this coming change, after 2012, is a totally different way of being. Once the wars and the demented state that creates them are out of the way, we can start learning again. Then I am coming back into a new body. This one is okay for what I have to do now, but it is limited. As my soul grows brighter, I am aware of the limitations, and of the damage done by the Draco-reptilian race to human genetics.

The regressive ETs were dumped here by another group of extra-terrestrials after one of their wars, and they are being kept here, because whoever it is does not want them *out there*. This is a very old war. I have been told their race is dying out because their genetics are imploding. Then it would make sense that they intend to take humanity with them. Our physical bodies are imprinted with their brutality. To heal it we need to inwardly change how we live. Only then can we advance.

For various reasons, different alien races want control of the earth. I suppose in the same way different countries want control of the Middle East. I would guess something powerful is going to happen here and maybe also because their worlds are dying, and they need our genetics to move and to survive. They are preparing to fight each other over ownership of the earth. Then there is the *Intergalactic Council* that is considering shifting humans into a new evolutionary phase. In other words: Contact!

I do not agree that we humans should accept extra-terrestrial technology into our world. Partly, because I know that we are capable of developing our own peaceful technologies, if we move in the right direction. Any meeting of two cultures will effect and change both, sometimes irreversibly. Rather than run blindly into the *look before you leap* contact-exchange, I feel that we have to step back and remote-view the situation... Thinking back to the Trojan horse, and the disaster that followed for the people of Troy.

In accepting and using this kind of technology exchange, we open ourselves to a scientific Catholicism, which is that, "they lead and we follow." Our societies then take on their ways, follow their restrictions and their developments and not our own. One way to destroy a culture is to present them with a view of the world (or Cosmos) that you wish them to have, and then manipulate that perception. If you are the designer of the technology, then the distortion of perception can be built into the design and the design can dictate the direction, or the levels of dependence.

When I talk of contact, I am talking about humans first getting to know *themselves*. Mahatma Gandhi wisely said, *"As human beings, our greatness lies not so much in being able to remake the world – that is the myth of the atomic age – as in being able to remake ourselves."* Present day society is breeding and *educating* a race of *slaves*. The gateless gate is the doorway to the soul, and the soul is a slave to no one.

The *Gateless Gate* in Cosmic terms is the ancient inherited *Lion Gate*, and one of the most ancient Cosmic star gates is *Ishtar*, or *Inanna*, the fixed star Spica at 22 degrees Libra. It relates to out of the body and astral travel, yet there is much more... something else that is much harder to define, that virtually no one has discovered. A few of us are related to this frequency by birth. More closely a birth of the spirit within the earthly realms.

Close to the body but slightly outside of the physical shell, hovering near the area of the heart (only higher) is a field of concentric circles like those drawn on the walls of Chaco Canyon. They are fluid like water, and yet they maintain their form. Their color is a deep blue, but it changes with our moods. This energy is fluid and yet it is also an impenetrable shield. It is a living star gate. Every human who is of the Sirian Lion race has access to this psychic field, and we all can move through it and into other dimensions of space time.

Those concentric circles may very well be a *Sacred Navigator* that aligns with the Cosmic star gates and allows us to move through the dimensions, as well as from world to world. Why would superior races from other star systems want to be teaching a secret government how to build certain technology unless we have something they need? And why keep it *secret*?

Perhaps it is our ability, or the star gate genetics missing link, an inner soul navigator that can do things they cannot achieve. That is the general information I get... that they kidnap humans who have abilities and use them to further their galactic empire. If we can focus on this Cosmic shield and alter our awareness to something greater, then we need no technology – we *are* the technology.

The difficulties facing us are a real, but the inner state is protected and sealed. It is a time portal. This is beyond out of the body travel, because it combines the holographic resonance of inner and outer space, of inner seeing and remote viewing. In fact I feel the people from the future are showing me a higher form of awareness, and I feel it is something *humans* but not all alien races can do. At least among the cat-bird people genetics this is known as the art of bi-location.

As stated in the ancient texts, jealousy drove certain older races to attempt to destroy the men on earth, because we are relatively new in this universe and bring with us new gifts and abilities. There are much older extra-terrestrial races who don't like it that we have been given some kind of ability that sets us above or at least beyond them. The concentric circle time portal connects to higher races, or higher universes that are beyond the regressive-alien hierarchy agendas.

Maybe we are the star children of the unknown sages of Cosmos. Which means destroying our world is not such a good move, if we want to learn true star knowledge. What we learn on our psychic journeys we can bring back into this dimension. That is why the others will not let us develop. In fact, I know some of them are afraid of us, because if we use our power wisely, we seem able to go beyond them.

Humans are in some way God's emissaries to this galaxy-universe. They have the potential to bring change, and can meet challenges that are not being met by the fixed races. God is shaking up reality. Just because a race exists, and has been around for a long time, does not mean it has an automatic right to continue. Therefore, if we – the soul inhabiters – do not apply ourselves soon to practicing the art of balanced living, we are too dangerous to have around. The word *extinct* comes to mind.

I feel a much larger universal change is on its way, and this change is not confined to the earth. What if the opening or activation key of the lion gate or star gate portal in the sacred sites, opens or awakens this localized shield in humans? What does that shield connect with? I think it is the life-cross or Ankh, within our DNA, i.e., the soul – that which the regressives do not have and can never get. We are in sight yet out of reach. This is our advantage. Let us play our trump card now!

Zones of Passages

"I must not fear. Fear is the mind–killer. Fear is the little–death that brings total obliteration. I will face my fear. I will permit it to pass over me and through me. And when it has gone past I will turn the inner eye to see its path. Where the fear has gone there will be nothing. Only I will remain."

~ Bene Gesserit: Litany Against Fear, Dune

It is a mistake to look for or focus on one single point when the location points of events exist as non-local evaluations over lengths or spans. What I mean to say is that you cannot understand singular events in terms of focusing on each event, but you have to understand the span of the event. Life and the events contained therein move in complete cycles. Which means if you set off an event in a certain astrological alignment, the effect will move across the span of the cycle, like ripples across a pond.

The pond is the containment of our physical awareness, our experience of this world. The time it takes for the ripples to cross the pond in physical terms can be years rather then weeks or months. Applied magic can alter the course of events; if the psychic mass is not so messed up that it singularly is caught in the event. Once the psychic mass is caught in the event, the future plays itself out according to the mass psychosis rather than the use of the intelligent mind.

The point-to-point energy of events that will evolve in this aeon, are overwhelmed by the length of the time period within which the events takes place. Cosmic mind moves in cycles and does not respond to individual local points of activity, which is why it makes sense to tune into Cosmos, rather than into the singular events.

We tend to consider each incident, and focus on the location and moment of a single event called memorable or catastrophic; for example, the JFK assassination, the landing of the first man on the moon, or 9/11. What you do not yet fully understand, and sadly, what the system managers and media manipulators use to their advantage, are the movement of cycles of complete time zones.

Our day to day lives are moving through, and exist due to, an invisible ether. Out of the ether physical life is born as molecules, and the molecules create matter. People, all of us, move through these *ethric time zones*. For most people it is the singularity of an event or occurrence which they remember, because most are unaware of the greater body of water – ether – through which we all swim.

We are in this present mess largely because, a long time ago, a manipulation of local-events was set in motion to alter the time-zone of a particular *Transit* embodying our human earth experience. The transits are like tension lines holding together an energy bridge intended for us to cross. You could say they are the tendons holding together the collective experience of mankind.

Elite *occult* groups, seeking to manipulate the mind of man, set intrusive events in motion into the ring cycle of our reality. These groups are not seeking to manipulate *single day* events, but are applying their fragmented knowledge to manipulate complete *Time Zones*. This is a secretive form of mind control, or behavior control in terms of millions of people across the planet, across time.

Only because the masses live according to this unnatural manipulation of cycles and events do we live in such psychic poverty. The time lords teach us how we can alter the natural time line for ourselves, creating balanced lives, and manifesting order out of chaos. To alter the cycle we would have to actually live the way of balance and order. The Japanese called it the Way of the Tao, or Zen. For me it is the *Zen of Stars*, as Cosmos is my guide.

The manipulated events that have succeeded in allowing the *shadow people* to colonize the world are manipulations of *Zones of Passages* or *Passages of Time* through violence and fear. This is the secret, or *occult knowledge*, elite groups have made use of throughout the ages. At the same time, this fear based system is coming to an end, as planet Earth morphs into a transmutation of metal into light.

If you consider how the knowledge of scientists such as Kepler and Copernicus were suppressed, or how the work, experimental notes and research of Nikola Tesla and Wilhelm Reich were destroyed by the American government. If you look at how the Native Americans were brutally crushed walking the *Trail of Tears* or at the *Massacre of Wounded Knee*; these events are no different from the JFK assassination or 9/11. They were all carefully planned horrors calculated to create precise damage to each zone of passage, and to damage the people passing through the zone or the *Transit of that time*. Consider carefully the word *Transit* and apply it to your consideration of the *passage of time*, or *passage to peace*.

Time moves in transits, as zones; and most important as zones of awareness. Dream of these zones and you will find the answers there. Ultimately humans will discover the solutions through their *dreaming power*. When we dream we naturally pass through zones of passages. We are now at an age where we need to learn to dream awake.

All humanity collectively travel together, "the passage of transiting through zones." The invasion of Iraq and other countries, and the approaching chaos still to be lived, are part of the misapplication of Astrophysics. The shadow leaders have learned, or have been taught to manipulate key transit points in order to effect the collective transition through each time zone. This is why their manipulative actions are so devastating for the whole planet as well as for the individuals involved.

Duality is the key. Cosmos created duality, not as a division but as a means to bring space into all that is. Duality is simply the pillars that hold reality together as time and space. Just as duality exists in all things physical, and that which can be measured, so it exists as two pillars standing between the zones. That is why I named my web site *Passage11*. They are the dual pillars of physical and spiritual existence, the Cosmic twins.

Look at the ancient structures in Greece, Samaria, Egypt, India, Mexico, Tibet, Easter Islands, Stonehenge, or the stone tablets of the Sumerians and Egyptians. Look carefully at the figures on the walls of the so-called tombs, which are in reality powerful places of Initiation and Knowledge.

As the opposing elite groups fragmented and fought each other, they hid the knowledge and they called the sacred places *tombs*. I call them the places of buried knowledge. I feel they were, and still are, transmission systems to other worlds; and we will activate them at some time again soon. I see star gates appearing as points of entry into each zone, and as we search for answers we have to look broadly at the complete passage of time zones and not only at the point of entry. Tolkien wrote in, Lord of The Rings: *"One ring to rule them all."* The most obvious consideration is that he is talking about a metal band worn on the hand as a ring, the one ring of power through which Sauron would rule all the world.

Over the past years there have been suggestions that a particle accelerator ring is needed for time travel experiments. What is a particle accelerator used for, other than the reasons given to the public? Preston Nichols says there is a particle accelerator beneath Montauk, Long Island and the worlds largest accelerator is at CERN Geneva.

In "Pyramids of Montauk", by Preston Nichols and Peter Moon, they write: *"What appeared to be a large circle cut out of the foliage was identified as a particle accelerator by my friend Danny, a nuclear physicist."* They go on to say, *"Bending time is not the only use of the particle accelerator, it is also used today as a particle beam weapon."*

Peter Moon also discovered in his research that a large particle accelerator is used to feed smaller ones. It makes me think how Sauron gave the seven rings of power to the dwarves and nine to the kings of men, but the one ring ruled them all... One ring of power feeding smaller ones, and when the one ring had been destroyed the smaller ones also lost their power. The development and use of particle accelerator technology gives new meaning to Tolkien's work: *"One ring to rule them all... One ring to find them... one ring to bring them all and in the darkness bind them..."*

Few people know that CERN was the birth place of the World Wide Web, which began as a project called *Enquire*. However, the web-project at CERN was never to be released into the public domain. Timothy Berners-Lee developed the protocols for the first web browsers and later the world's first web directory. He made his work freely available to everyone with no patent and no royalty free so that all could benefit and use the system he developed.

In Berner-Lee's book, *Weaving The Web* he makes it clear that, *"Computer scientists have a moral responsibility as well as a technical responsibility."* The nature of this development and the collective world wide benefits are one example of the correct and balanced use of a zone of passage. The man who had the technical and artistic skill to develop the first web browser also had the integrity to make sure that its development would continue free of restrictive royalties and patents.

I feel and see a positive outcome, although hard times will have to be faced, as it seems the more regressive aspects of mankind do not want to step aside to allow intelligent cooperation and diplomatic world peace forums. There is something here on earth that desires power over all life, as Tolkien so perfectly showed in his trilogy. Humans would have to stop being seduced by its promise of riches, to understand where they are heading if they agree to the prison walls.

At the *WingMakers* website, Dr. Neruda talks about the coming development of a totally new technology. Of this technology Neruda says: *"All I can tell you is that it's related to the Internet and a new communication technology that the WingMakers referred to as OLIN or the*

One Language Intelligent Network. If you read the glossary section that I left behind, you'll see it referenced there. The WingMakers seem to feel confident that the OLIN technology will help create the global culture through the Internet. This incidentally is consistent with prophecies that the Labyrinth Group was privy to dating as far back as 1,500 years ago. Of course the enabling technology wasn't called OLIN, but the notion of a global culture and unified governance has been predicted for many centuries."

Is it possible that longer living inner-world reptilians sought to control planet Earth and lost their control in some kind of war? That these beings are looking for a way to use humans to regain power on earth and fight their enemies? Is the Sphinx on Mars part of an ancient network of passages between the planets? The photo of the Sphinx and the huge pyramids on Mars, are parts of a mystery being deliberately hidden from the people on earth.

Finding the *Zones of Progressions* will bring even greater understanding of the timing of events. Time is measured in ages, thousands of years; the famous Mayan long count. A friend of mine once said that I have a memory worthy of the Mayan long count; well, let's see what my long count tells me when we get further into my discussion with *The Master of Light.* He says that in the age of knowledge we will leave fear behind us as we begin to ask the right questions. The wisdom of the ancients is hidden from the eyes and ears of men, and will remain so until we learn to use our power for good.

Transmigrating Emanations

"I swiftly become their savior,
From the world that is the ocean
Of death and transmigration,
Whose thoughts are set on Me, O Arjuna."

~ Lord Krishna, The Bhagavad Gita

It was late at night and the streets were empty. I was standing outside the castle walls and thought I was awake, when suddenly the castle walls were standing outside of me and I was met by my friend, *The Master of Light*. I did not awaken from the dream, the dream continued. He led me towards the lower stairs and I followed.

I did not stop to think whether I was awake, no more than I would if I were awake when walking in my daily life. The meeting was just as real as our world, if not more so. We descended the stairs and approached the natural stone rock close to the portal wall. The long room was not dark; it was brightly lit with an unearthly brightness. The light did not disturb me, I found it pleasant and still I did not stop to think or consider that I was *dreaming*.

The door between the worlds was open and active. Their world shone into ours and the granite stone took on an opaque hue. For a moment it felt as though their world was real and our present world a shadow cast on the walls of time. I followed *The Master of Light* through the doorway, the floor changed from stone to light and I was suddenly falling out into the infinity of a starry night sky.

There was no *Master of Light*, no Chillon castle, no planet Earth. I was alone. At that point I figured out either I was dreaming or something was wrong with my hold on reality. Actually, flying in space felt natural to me, I have to admit. I no longer had, what I thought was my body when I had entered the castle walls. I seemed to be a floating nebulous mass, a mass of light or intelligent awareness.

The one thing that was strange is that it was not *me* who was directing my out of the body space travels. Something seemed to be drawing me into space and towards a specific destination. I sensed that the time lord was somewhere at my side, but I could not see him.

It was not long before I saw my port of call, my starbound destination. Below me was a small planet, similar to the earth; but I knew straight away it was not planet Earth. All information came to me psychically, as though I was connected to a Cosmic intelligence or a Cosmic psychic Internet. I knew things without knowing how I knew or where the new information came from.

I was obviously floating down to this planet for a reason, and so I paid attention to understand more. It very soon began to look like I was back on earth, because I was coming close to the unmistakable scenes of a war. If you think now that wars on earth are bad, which they are, then this one was far worse.

In the scenes below there was death and dying everywhere. The surface of the planet was ravaged by weapons we cannot imagine. I immediately wanted to pull back and hide. I felt the danger and instinctively searched for a way to protect myself. I wanted to get away from the dark horrible feelings of the place. The physical destruction was bad, but even worse was the dark feeling of hatred. In fact, the hatred was more toxic than the scenes of death, as the death was emanating from the bad feelings and bad thoughts the people on the planet had towards each other.

Powerful groups were struggling against each other for control of the planet's resources. Maybe the scene sounds familiar to you? From above, in the psychic realm, I could see the hopeless moves each group was making, with no success. This was to be a fight to the bitter end, and it did end – with the destruction of the planet.

These humans hated and feared each other so deeply that they could not longer function as *humans*. If there had ever been diplomacy or agreements of cooperation, the fact was that it was no longer an option for the people of this other world. As I watched them, information came to me out of the psychic ether that the people had crossed some irreversible line of insanity. Even if they wanted to, they could not cross back; such was the level of their insanity and hatred.

I watched the fighting which had, at some point, become fighting for fighting's sake. The resources and the infrastructures of the societies on the planet below me had been ripped apart. The military of each opposing group had the greater part of the economic resources, and the people had the least. In some way this hardship only strengthened their resentment and their bitterness. No party was going to give up its share or give up the fighting. The planet was doomed.

The worst part of what I saw is that it was clear no one cared. I thought about our world and at least some people do still care. But I realized also the parallels between our world and the world I was looking down upon, it was the same greed and desperation to control the resources of the planet, at any cost. If the countries on planet Earth took on this role of ownership at any cost, then the results would be the same as the ones I was now witnessing.

From my psychic viewpoint I watched the total destruction of another world as though I was looking through a holographic window of time. The scenes were horrific and not something any sane being would want to live through. It was clear to me that the people on this other world were anything but sane.

I thought that was the end of my Cosmic lesson and I expected to return home. I was already feeling not too well, having seen this other-world nightmare, and I wanted to get out of there. The next thing I saw was, in some ways, even more disturbing than the planet's end.

The physical planet was a lifeless, burnt-out shell floating in space. Nothing survived, or so I thought. Then, from my psychic spirit view I saw some kind of movement in the space around the burn-out atmosphere. What exactly was moving, I could not see, but I could sense it. The movement was something like a black or dark psychic cloud. I knew it was the entity-like thoughts of the hatred that had polluted the minds of the people fighting the wars.

The psychic darkness began to search for life and found none. In fact I wondered if this dark force had in some way polluted the hearts and minds of the original inhabitants of the destroyed world. I then watched in horror as the dark cloud of psychic hatred left the lifeless shell of the destroyed planet and headed into the depths of space. In my heart I knew it was headed for earth.

The other race that had destroyed themselves and their own planet were much older and more advanced than we are now. They had been more technologically advanced and in some ways more brutally regressive. The psychic force of darkness came into our solar system and arrived at the earth in some distant past age, perhaps even at the time of Atlantis.

The cloud of fear and darkness was searching for an energetic life force to feed from in order that their entity thoughts could continue to exist. But the energy they needed had to be of a certain vibration. It had to be the energy of hatred, the energy of fear and suffering, destruction and war.

They made a secret alliance with the dark priests of Atlantis to live through them and be a part of them, and manifest their psychic power through them for control of the earth and the earth's resources. Secretly the alliance was made and the dark force passed into the psyche of man. It became a hidden part of men, unless it revealed itself in act of hatred and revenge. The only protection against its dark malice was gentleness, understanding and love.

The Tribe of Light saw the darkness take hold of the wayward priests, and they began to take action to limit the destructive power being unleashed on the earth. Atlantis sank beneath the waves and the dark priests lost their power base.

Although the parasitic dark force was haunting the lives of mankind, it could not take control of sovereign minds, people of good heart, and it fled from those who know and emanate love. The other-world psyche had sought to emanate from its world and transmigrate into the bodies of earthmen. The battle for the mind of man had begun.

The dark force was not destroyed when Atlantis sank. Until today it seeks to manifest and feed off those humans who live its ways, creating violence and wars, manifesting hatred and fear. This would be a spiritual war, played out in the physical arena of planet Earth. The only way man would survive is if he chose to live the life of the divine, to escape to *higher ground* and leave the dark force to die to itself.

I knew then that humans on earth were unwittingly repeating the destructive cycle recorded in this psychic parasitic other-world force. I saw the different doctrines and dogmas take form, but in the end it was all the same hatred and greed. The alien psyche could not take physical form or incarnate into the physical-material realms, but the greed and desire for the material was always present and the hatred of mankind even more, out of jealousy.

Man's desire for *power* was his downfall. The dark force was soon spreading into the fragmented and conflicting groups, and in pockets all over the earth the insane other-world conflict I had witnessed was taking form in our world. Richer and more powerful nations were transmigrating onto smaller weaker nations, taking control of the resources and destroying the infrastructure of the host.

The *religious-political* military solution is becoming a nightmare as the people on earth play their roles, in a badly edited script, in which the house burns down, and the people watching also get caught up and consumed by the violence. The fear had entered our world as a visitor

223

from another space-time, and yet humans were living the insanity as though they enjoyed eating the toxic food. A food that was neither feeding nor nourishing them, it was poisoning them.

The only way to protect yourself from this force is from within. The dark force feeds from negativity, conflict, fear and sorrow. If you don't feed it the force will fade and die. It has no life structure of its own; it needs human insanity if it is to continue to exist. Be sane and be safe. Place yourself within the emanations of the Divine and out of reach of the emanations of fear. The *golden heart resonance* of the Divine is the monatomic gold of the spiritual realms.

I returned to the castle alone. The lower dungeon was completely dark, and the time portal was a wall of cold stone. Nothing stirred, not even a mouse and the outside world seemed far away and lost in a sea of mist. At the end of the long dark room I could see a faint but growing light, like the light at the end of the tunnel. I knew the mists would eventually consume the dark thoughts and all negativity would return to the ether. The approaching light was something new. For the first time I felt I was the time lord, then I woke up in my room.

Transmigration – the act of removing to another country; the passage of the soul after death to another body. *Transmigrate* – to pass into another body or planet.

Designing A Sane World

I-Magi-Nation Is Survival

We Came From The Stars

Exploring Human Migrations

Inter-Dimensional Schools of Light

Contact With Trans-Human Souls

Cosmic Age of Knowledge

Strategy of Saturn

Future Earth Security

I-Magi-Nation Is Survival

"I knew, of course, that trees and plants had roots, stems, bark, branches and foliage that reached up toward the light. But I was coming to realize that the real magician was light itself."

~ Edward Steichen

The greatest gift and best tool humanity gave up in this past century is the power and wealth of our imagination. Rather than being encouraged and developed in young children, imagination is subtly but purposefully stifled. Teachers today, as those over the past century, have lost this gift. They have become robots to the system, and so they show children a miserable world without imagination. They teach stupidities at best and untruths at worst.

However, in my consulting work I have seen that fortunately at least some of the psychic and *Indigo* children, and some young adults do not buy into the nonsense. Thirty years ago, I rejected it myself. These new young leaders, teachers and healers are seeking guidance from the light; they know we cannot carry on in this destructive way. But, the true idea of what it means to use the imagination in times of survival has also become a target of religious, scholarly and political disinformation.

How we will make it through the changes is an art form yet to be developed. The earth, its peoples and its nations will have to face the truth of the actual and secret brutality of their centuries old behavior. I advise that people do not try to reorganize the illusion, as the transiting planets will not all benefit such a move. A truly psychic mind would not destroy life, but it would be part of it; and indeed, create a new and sane life. Enlightened Minds use *Imagination*.

Imagination has been presented to us as a sign of eccentricity, as a lack of discipline and lack of development, or worse, even as a sign of retardation, an escape from reality. It may well be an *escape* into a better and saner world, and if that is so then let us use it to get out of this present mess. Houdini was the greatest escape artist ever known and he used his imagination to show us how to achieve the impossible.

When I was a little boy, I imagined things and asked questions, and I was told not to *count the clouds*. I was instructed to: *be practical*, and *not dream*. Hence my dream of becoming the greatest chess player had to be shelved as an unrealistic option for my life. I am certain many other kids had the same experience and gave up their dreams. So, I pretended to be practical, and I kept my dreams alive by studying other matters, until I came to discover the *Zen of Stars*.

People who use their imagination are apparently not in touch with the *real world*, or so they are told. They are seen as wasting their lives *dreaming*, but never achieving anything *real*. Great thinkers of our time are sold to us as the superior minority, and we are apparently not in their league. Small people do not dream, they *work*... But who can stop you dreaming if you refuse to cooperate with these soul destroying rules and regulations of the matrix?

Developing imagination is looked upon as a danger to society, or as a form of insanity, because the system needs robots that conform, not intelligent sovereign individuals who are in command of their own mind and are actively using their creative talent to bring about order, peace, sanity and cooperation.

Without imagination you obey. Like a robot, you do what you are told and do not question. I say that inquiry is imagination, which is vital to survival. In other words, imagination *is* survival, whereas all else is death. Our society is now being threatened by so-called artificial intelligence, so what I describe here is extremely relevant if we are to survive.

The greater part of the human population does not, and cannot fit into this society – into this world – the way it is structured now. Most of the people on this planet are poor, struggling without basic food and clean water to drink. Most humans on the planet are struggling just to live. Their children are suffering quietly and the little the colonial structure has to offer them does not resonate well with their genetic intelligence.

To succeed within this society you have to give up something precious: your freedom, your humanity, your creative imagination – and worst of all, you hand over your love to the robot masters. By doing this you forego your existence and destroy your future.

Today's social structure, its patterns, and its rules are self and planetary destructive. Modern society disrupts human intelligence and the inner knowing of the heart. To be allowed to participate in social groups and to exist in this world, each individual has to agree to rules that are set up to damage individual growth and thus damage society itself, and yet no one is to question this.

This does not have to be so; we can change the future

through changing the now – using our own unique powers of *I-magi-nation*. Many of the older cultures are struggling to survive on the most primitive levels, and they are being kept in this position to prevent them from growing and developing in a new and more balanced way. The shield that actively protects us against this dehumanizing force is our own and our collective creative imagination, which is the power to dream and to realize our dreams. It is time for each of us to reclaim this most important shield, because it is also our guide and it works in contact with the heart. The future of humanity is the path of the heart. Imagination is the key to opening this path.

Imagination is also part of the power generator that creates an aura of freedom around us. This means that outwardly you may have to work within society to feed yourself for a while, but inwardly you are far away and out of reach. Using the inner shield of imagination, you remove your psyche from the limited field of social structures, thus withdrawing your energy. The structure of society seeks to destroy your imagination, because it needs you to *believe* in its authority in order to make it work. If inwardly you have removed this belief by developing your imagination, then at that point, something new will take place.

If your imagination is crushed to a tiny flickering flame, however, and your whole life force goes into believing society has power over you, then society *will* have power over you. Which is only so when you make it your current reality. The truth is you can choose which way to go, and know that society will experience its meltdown in a few years, which is life's way of challenging you to act now.

Imagination empowers us in a unique way, when we invest some effort into developing it. It is possible to actively participate in society and at the same time direct our inner power to imagining a new and different world. People are mind controlled to believe there are no alternatives to cars, airplanes, or wars, for that matter. The mind control is part of an agenda run by elite controllers who have no interest in the well being of people or society at large. You are made to believe, for example, that there are no alternatives to oil, gas and nuclear power. But if you delved into your imagination, you would discover that this is wrong.

Coral castle was built by Edward Leedskalnin, in the 1930's in Florida, USA, with the power of and through his *I-magi-nation*. This small man was able to carve and move massive blocks of Coral rock, some weighing as much as 30 tons on his own, and no one knows how he did it. He wrote, in 1945 that...

"As I said in the beginning, the North and South Pole magnets they are the cosmic force. They hold together this earth and everything on it, and they hold together the moon, too. The moon's North end holds South Pole magnets the same as the earth's North end. The moon's South end holds North Pole magnets the same as the earth's South end. Those people who have been wondering why the moon does not come down all they have to do is to give the moon one-half of a turn so that the North end would be in South side, and South end in the North side, and then the moon would come down.

"At present the earth and the moon have like magnet poles in the same sides so their own magnet poles keep themselves apart, but when the poles are reversed, then

231

they will pull together. Here is a good tip to the rocket people. Make the rocket's head strong North Pole magnet, and the tail end strong South Pole magnet, and then shut to on the moon's North end, then you will have better success."

The above extract is from, *Magnetic Current* by Edward Leedskalnin, one of several booklets that he wrote. The existence of Coral castle is one example of the individual power of imagination, of a man living true to his dreams and manifesting this power into the physical world.

Today, a small number of inventors and creative thinkers are struggling to make free energy and new health empowering innovations a reality. They are moving against the tide of this sick and imprisoned humanity. They are the apprentices of imagination. If people would imagine and explore new ways to live, untried methods to build houses, and alternative solutions to distribute power, it would all come about. But if the mind is in resistance, if it is paralyzed by the controllers, it is much more difficult to make a breakthrough.

All we have to do is change our minds in order to create change. Yet that is a tall order in a mind-controlled society where people live the lie they are told. People in the *civilized world* had to be convinced that it was necessary to invade, for example, Iraq using lies and deceit. Once the action had been taken it did not matter that the truth was different than what was claimed. In this case, there was no reason for a pre-emptive attack, because no weapons of mass destruction have been found in Iraq... but it was too late, the damage was done and the controllers had struck another victory.

It is the same with using oil and nuclear power as a source of energy. People are made to believe if you take away this source of energy, society will be unable to function. Fear maintains society as the master of the people, instead of the people being the masters of society. If more people were to develop and use their imagination – if they would just think on their own, without someone feeding them a pre-established belief system and if they would ask questions – they would realize there is something drastically wrong. How can society be the master of people's actions when there is no society without people? Imagine it is the other way around and it will come into being. People will become the masters of their own social order.

Imagination is an important doorway to the part of the brain that can view the future or the past. The most destructive deception used by the establishment is the lie that the brain can only perform one function or concentrate on only one thing at a time. This deception is aimed at destroying our capacity to use our imagination.

The psyche or the psychic is the most important element in creating anything humans develop or use. Tools, shelter, clothing and transportation all originate from our use of the psychic, creative and imaginative aspect of our nature. We do not bring all this about simply through thought. Imagination is not thought, and the psychic is not thought, although both can use thought as a tool to carry out or develop their aims or insights.

We generally accept everything, and that is wrong. We rarely imagine that we can change things or improve a situation simply with the power of the mind. Much of our lives are lost in negative thoughts and fears. But imagination, when used properly, can overcome negativity and certainly erase fears. If you simply imagine a situation to be other than it appears, then this use of imagination already changes the chemistry inside the body and the brain. Just blindly accepting what is in front of us is a failure of imagination. It is also a failure to realize that the real power resides inside of us and not outside.

Everything we have built lacks imagination, and this creates a feeling of helplessness and a fatalist's perception of, "what's the point" – when the point is for us *to change it by using imagination as an aid to developing a new world.* In your imagination you can fly anywhere in the world and zap beams of light or Orgone power into the darkest corners. In your imagination you can see the darkness returning to light. The limitations to this are only the limitations you set on your own power to imagine. Light shields can be created around the physical body, using imagination. Healing sessions can be organized from and through the power of the mind.

Personal conflicts, problems at work, or family confrontations can be transformed using the light of imagination. You can create your own world with your own mind and transcend those unwanted situations, conditions and issues. You can dis-empower past negativity and delete negative events, using your mind. This is a self-empowering form of visualization, and it works.

If you work in an office, you can do your work and at the same time you can imagine yourself: swimming in the sea, standing on a mountain or even flying to Mars. Which are subtle ways to remain healthy and sane. There is no limit to your imagination. The body and mind receive the good energy of whatever you imagine. But most important of all, you deny society power over you or your mind. The structure of the world today has a negative effect on the natural human system – that natural system is the oldest genetic awareness within us.

I spent most of my time at school using my imagination to go for walks in the mountains, or sitting on the sand and looking at the sea. It relieved me of the stress that we call education. I was happy. No one could control my thoughts, so I was free. The teachers had no power over me. There are no set rules to developing your own power of imagination. It is your own creative Zen garden of light.

Each individual has their own unique energy shell, and the development of imagination is just as unique. It grows out of who you are, what you have learned and what you need to learn in order to grow further. Imagination is the tool we use to contact the spirit world or our guides. Naturally if the thoughts are negative, these are not right guides, but entities who wish to make us go in the wrong direction. Remember, every move effects you. So, it is better to use imagination to overcome negative thoughts rather than be overcome by them.

If ten percent of the people in society began to use imagination as a tool to change society and our energy supply source, things would be different. Instead, we simply complain and wish: *"If only..."* Only when each one of us has the power to say how it is done and not *if* – will we use this unique psychic gift to transform our world, and move forward to create a new world for ourselves. The power to imagine something different will create something different. That power has been largely blocked by fear until now.

The truth is: no one can control what you think. No one but you knows you are swimming in the ocean in your mind while you are working at your computer, or picking corn in the fields. I am talking about using the mind expansively to be perfectly aware of the tasks of this world and at the same time be perfectly aware of the power of the psychic world. This ability of the mind is also known as *Psychic Recall* or *Remote Viewing*. Expand its skills and you may be surprised at how you are better able to function in the day to day physical world, be aware and master that world.

Our brains do not work at optimum potential because we do not use them in imaginative ways. Throughout most of our lives the instrument between our ears remains switched off and largely inactive. Imagination is the key to self-empowerment. It is also the single most valuable tool of self-protection and thus of survival. It can be a lot of fun too, because we can create new solutions to old problems. The more you imagine, the more you know...

You are not born into the outer world; you are born into the inner world inhabiting the outer world.

We Came From The Stars

"Knowledge was inherent in all things.
The world was a library... "

~ Chief Luther Standing Bear, Oglala Sioux

One of my earliest memories is of the stars. I was always aware that we are star beings, star children. Even when I was very young I would look up at the diamonds of light in the heavens and ask the beings up there to come down and, *take me back home*. Which ended up to be a long discussion as to why this could not be done... At present.

Michael Cremo says, "We did not evolve up from matter; instead we devolved, or came down, from the realm of pure consciousness, spirit." Which is the feeling I had as a kid, looking back to my place of origin – the stars. He bases his response on the findings of modern science and on the study of world's great wisdom traditions, including the Vedic philosophy of ancient India.

Cremo proposes that before we ask about the origin of human life, we should first contemplate what a human being is. He asserts that humans are a combination of matter, mind and consciousness, or Spirit – which translated to energy, since matter is also energy.

I would take this one step further. We are all unique beings: light beings, human beings, alien or extra-terrestrial beings, evolved and devolved beings. We have ample scientific evidence showing how a subtle mind-element and conscious-self can and does exist independent of the physical body, and how this fact of life has been systematically eliminated from mainstream science through a process of *knowledge filtration*.

"Any time knowledge filtration takes place, you can expect a great deal of resistance, criticism, and ridicule when it is exposed and challenged," says Cremo. Which is why Immanuel Velikovsky was attacked by science in psychotic ways. Velikovsky's revolutionary writings totally changed the world of science and the esoteric history of mankind. His account of events in *Worlds in Collision*, and *In the Beginning* are masterpieces of Cosmos.

Keeping this in mind, and looking beyond, into the approaching times ahead of us, we are headed toward a most amazing experience. Our entire environment consists of signs and symbols. I asked at the beginning: "What is real?" Because we see reality in a defined way, it does not occur to most people that they are weaving an illusion – a self created mandala, and either this mandala is balanced and whole, or it is fragmented and corrupt.

Our entire mental landscape consists of these symbols. The hologram of this present reality is made of agreed upon symbols, and those agreements are straining at the limits that have been set upon them. To build consensus for the sake of peace is to agree to disagree. You have to agree to the reality being presented to you, which is why I say "free your mind!"

Krishnamurti said it even better in *Freedom From The Known*, when he said: "Thought is time." Everything we are about to experience will be a wake-up call of absolutely unimagined proportions. This is going to look extremely crazy for a time lasting several years, and during the chaos many will cling, for dear life, to fixed meanings of favorite symbols and beliefs. Better let go now to keep from going mad as new icons of destiny will map reality.

Lifetime after lifetime, each individual soul or being can decide: will it move to the lower vibration of the Cosmic lesson, or will it transmute lead into star gold? Once transmutation has been achieved, the being discovers there is no higher and lower; there is only what Krishnamurti and the sages call the: *what is* type of reality.

To discover this requires energy, and how we live now either decreases or increases the life force energy of the radiant body. I consider the magic ring is our own DNA ring, and our connection to spirit, light, or the divine. Our mind connection and imagination can and must activate solutions for what I term the contactees world of intelligent survivors.

The self or ego, matter, money, power, sex, religion, ambition, greed, pride, wanting, fear, etc. and the world of *beliefs* are on their way out, because they are useless and counterproductive. This dysfunctional approach to life will be replaced by a scientific *knowledge* of the spirit. Belief is also the corrupting factor in human society because of the way it is used. It rewards those who bow down to the *golden calf* of ego and materialism, and seeks to destroy those who don't.

These people don't realize yet that one day their money and material riches will be taken away from them, partly though circumstances we face, partly through earth changes. That day is dawning soon. Then, what will happen to the worth of the ego and its self-serving materialism and service-to-self agendas?

If we work together and create light technology or non-technological orgone accumulators and other healthy energy based solutions, we will be well on our way to getting rid of the regressive alien type reptiles and entities feeding off mankind. This move can be done peacefully; it only requires a shift in the heart coherence. In reality, we know the course is set. Until 2017, there will be some destruction until we cross over into other dimensions. By *other dimensions* I mean the ability of the human being to shift realities and live differently than he is doing at present.

We are the stars, and the universe is our home. There is a non-technology radiance that completely blasts an area free of DOR (depleted orgone radiance). If directed toward entities or regressive type aliens, they will vanish quickly. The orgone generator built by Wilhelm Reich, is an instrument that is plugged into the ether. It is said to heal dying areas and can create cloud naturally. I would say that advanced humans are also orgone generators, and can create their own balanced resonance that also heals vast areas.

Using and directing your own intent, and focusing on the solution is more than prayer, and much more than mere meditation in the void. This is about unbending concentrated focused light beams of thoughts in search of signs, transferred via telepathy and acting like a powerfully amplified signal. The signal is focused to be caught by whoever can and will help. It is called *making contact*. What Reich discovered for orgone you can also do with the mind.

If an abductee – abductees are not the same as contactees – knew how to focus or *imagine* orgone protection, they would free themselves of that experience and quickly lose a few bad friends. Contactees are those among us who complete mind altering sessions of voluntary hyper-dimensional interaction in what I would term *contact centers*, or inter-dimensional ET schools of light, on their floating ships. My own light-craft vision is one example of my interactions with them.

Why do the controlling powers – governments, the church, special corporate interest groups with money and influence, and the media – want to repress our creativity and inventiveness? Why did they want to destroy, with a vengeance, the work of Galileo, Krishnamurti, Reich and Tesla?

The two are described as the forces of light and the forces serving the darkness. The dark force or negative entities feed off human fear, sorrow, feelings of loss, regret, resentment, etc. Therefore, in order to stay alive and do their work, they must terrorize humans, and in turn humans have to live in fear of them. Which keeps humans in a pretty brain damaged state.

I am sure we can create simple, non-electrical handheld orgone devices, or Tesla plates and use them to protect ourselves through the natural transmission of a radiant balanced energy. A client of mine, who is a technical wizard, told me that orgone devices would dissolve DOR and protect us from the dark entities. Jewellery loaded with appropriate light energy, for example, will protect us. The technology exists. We can develop peaceful and non-lethal DOR blasters and specific zappers which will clear an area of dark spirits, entities, negative aliens and underground reptiles.

It was also explained to me that the control system in place today eliminated the ancient cultures because these advanced civilizations were also contactee societies.

Anasazi cultures, the *ancient ones*, the Celtic and other nomadic civilizations were highly developed people who were in constant contact with the space beings, or other dimensional time travellers. Imagine a dark entity trying to force its way between you and those who wish to contact you. If we achieve Contact – and we will – the regressive ones will lose control of the earth. This is the spiritual and evolutionary battle that lies ahead, and your choices decide the outcome!

Which do you prefer – contact or control? The contact cultures were anti-control, for they thrived on a psychic science which some incorrectly call a *pagan religion*. These people had no religion, and the present civilization is the one of *pagan beliefs* disguised as worship of God. The fact is, these were extremely advanced knowledge and truth based cultures driven by Cosmic consciousness.

Those coming to us from the future are also working with us, because what we create effects them. They will inherit our mistakes. This may be difficult to understand because it relates to hyper-dimensional time travel and teleportation – but in essence, the way the light beings explain it to me, is that a Cosmic feedback loop of consciousness is ongoing.

If we create a particularly bad future, and mess up the planet, it will feedback into their loop and alter the light course of events in their world, potentially sending the whole time-field into a negative vortex spin – a spin they cannot control, because we created it and we are the ones to fix it. We are all one across the fields of time and space. We are all inter-connected through and beyond time and space dimensions as unifying field of Cosmos.

Many people are waiting for the elite to crash, or for

the Northern people to disconnect the elite, withdrawing the power to terrorize the world. It is possible that alien races are helping the secret governments to create a terrifying technology simply because we are coming to a time when humans will change and cannot be used by them any longer.

Experts and scientists, government people and military exponents will come forward and start to disclose what they know about projects they worked on; from alien contact, star gates, government deceptions and mind control; HAARP frequency experiments and the aimed for 2025 military control of the weather, the reverse engineering of crashed disks and the secret war being fought.

These discussions are still riddled with disinformation, but the information is coming out, and soon the dam will break and the trickle will grow to a flood. Everything will soon be in public domain, and the real action can begin. The ancient magi knew that the future is the extended now. It is a race against time, in this war of the magician. The alternative futures are primitive and to be avoided at all costs.

A few world survivors and contact cultures will start again, anew. We will find and protect the star gold... The race against time ends with the ending of time as both karmic response feedback and time accelerate. At that point comes a new physical vibration. We will see what that entails. A day morphs into an hour, and then a second. One day, we leave the polarity of time behind us. *The Master of Light* looks in and converses with me from beyond time, as we both come from the timeless stars.

Exploring Human Migrations

"Let us put our minds together and see what kind of life we can make for our children."

~ Sitting Bull

Mother Earth will sometimes shake a civilization off her back, and this is exactly what will occur during the next few years. We must decide if we want to create a bridge to the next civilization or perish with the existing one. This will be your own decision, and possibly a matter of personal taste.

We are going to witness this time of change soon, most certainly by 2011. Communities will need to relearn simple skills, such as navigation by stars and survival in unfamiliar environments. Urban people who are unaccustomed to living close to nature will probably not survive during these times of turmoil, civil wars and upheavals. Many people will leave the cities, to stay out of the way of trouble.

Those who are unaccustomed to living simply and naturally will need time to learn, yet time is running out. For some it may be too late to shift gears mentally or physically at the point when the earth's immune system kicks in. The adventure will be exciting, but not for the faint of heart. Unhealthy lifestyles have sapped 80 to 90 percent of the human population of their vortex energy. Spiritual attunement and personal changes will be necessary. The new DNA program will be called *shapeshifting*.

Leaders of the shape shifters will be the children and resilient ones who stay young enough in mind to move and adapt. The idea of clinging to a "homeland" or shell is ridiculous and unrealistic during a period when Saturn, bringer of form and controller of structure, will check into the viability of Pluto and Uranus by 2010, as a cardinal cross forms in the heavens. Saturn shows the weak link and brings the new form. Homes, nations and alliances will tend to break up under this influence. Peace treaties will be negotiated by tribal leaders world wide. Russia will do well in that phase.

Magicians and migrators are best when in constant motion, yet they remain quiet inside. Like moving targets, they will form new groups and teach the people their ways. In Summer 2010, Uranus and Jupiter enter the star sign Aries at the point where the Celtic new year begins every year. The Sabian symbol for Aries one is suggestive of the new time: *"A Woman Just Risen From The Sea; A Seal Is Embracing Her."*

Neptune at that moment hovers in the Aquarius-Pisces cusp zone, which is the vernal equinox area meeting the change of age zone: *Changing from belief to knowledge.* This is the guided time to do the right thing.

2010 and 2011 are extremely innovative years where only untried works and new trails are to be blazed. A remarkably constructive tension will be in the air until 2018, when Uranus moves into Taurus. I definitely see a completely new set of social values emerging. Great experiments will begin during that decade, pointing us to a new time. It will be the time of the new indigenous peoples rising.

The magicians of migrations will have to be trusting of this planetary aspect, and yet they must also remain impenetrable to deception by the controllers of society. They will be profoundly courageous and very firm, unable to be overwhelmed when this phase gets underway. Migration of people toward new and necessary resources begins in late 2019 with planets lining up in earth signs.

2020 is the time when I do foresee a shift of ages. Here I salute a colleague magi, one of the first modern tribal leaders of the earth in this new age: *"I invite you to be proud of the indigenous people who are the moral reserve of humanity."* From Evo Morales speech, as the first indigenous Indian President of Bolivia after 500 years of colonial rule.

Migrations follow the patterns of spirals. Pluto in Sagittarius until 2008 represents transformation of belief systems and ways of finding the truth. The planet of death and rebirth has traversed half of the sign of the archer since 1995-96. That is when the inner journey began. Now we are on the outer journey.

Many may concur with me that in 1996-97, parts of their lives became unglued or took at least some unexpected turns. Since that time, only a few have adapted to those unseen changes. Why is that so?

Some of you may have started to migrate into inner as well as outer space. Those who want to survive, without clinging to the old, will continue to see far ahead, reaching for the blue star. People have often been displaced and forced to move. Human beings have had to learn and re-learn new cultures and customs or adapt to subtle or sometimes sudden changes.

The ancient shamans were also explorers. Artistic visions, monoliths and carvings, ruins and temples are our witness those journeys. They are silent testimony to the great distances we have traveled, to what we have lost and what wisdom has been retained. Only few are capable of linking back to the common spiritual denominator of star magic, the secret code of the planet's mechanism that shows the way ahead. I came here to regain and redistribute this knowing, to prepare for the time when it will be necessary for our collective survival.

We are here to learn and teach others. We will remember fundamental skills as we become downwardly mobile. As a result, we will share resources rather than only look out for ourselves. Human and planetary resources are stored within our grasp, led by spirit and accessed through attunement with the planets at this new time. For many years I have been advising clients and friends to position themselves ahead of the changes. Only a few have acted on this advice.

After 2010, some nations will live in virtual self protective lock-down, with trading and travel almost suspended, next to impossible, or under control of authorities trying to impose order when chaos is the norm. It must be understood that the authorities fear nothing more than change, and they are fully aware that they cannot control those who organize themselves.

Only inventive and resourceful minds will be able to regroup, by declaring sanctuaries of relative freedom and by implementing functional economic systems of mutual benefit to all. There will be a growing psychological resistance to, and defense against, the obsolete super powers. Big business is nearing extinction. The economy of service to self will experience melt down. Small will be beautiful. Independence and individual liberty will become a prized possession beyond any money. Freedom of movement will mean financial prosperity. New skills will be learned. Migrations of magical societies will once again follow the pattern of spirals and infinity symbols. Explorers during and after the time of the global war games will not move from point to point in a line, but rather in expanding and safe networking circles.

Methods of relocation and trading will remain concealed in order to be effective. Envoys will go forth, bases will be established and people will be taught how to create intelligent networks of survival. They will develop preparation skills in order to be able to deal with the unexpected.

Migrators will emerge unexpectedly, linking themselves together via tacit agreements and silent understanding. The world can be gained by peaceful interaction, but not taken by violent control. People's hearts will follow the leadership that resonates with peace and joy. Intelligent leadership needs to be assembled throughout the world and effectively coordinated to benefit the pockets of survivors.

Smaller communities will survive through benevolence and kindness, tolerance and justice, inner and outer peace. Nations without justice or virtue will perish regardless of their size. That is why the free thinkers will leave the structures early and regroup in loose form, shape shifting into new interdependent smaller societies.

Survivors will be those who possess the inherent virtue to move with the least resistance, like water. In practical terms this means those who want to survive must be able and willing to move fast and efficiently, to move freely and remain invisible. Either we let go off the old non-functional ways, or we perish.

Detachment is the way to achieve this paramount state of inner freedom. The new *Migrators* will have these virtues. People will be willing to work with those leaders who have gained respect by their sound character and ability to help others survive. This authority will require neither laws nor constitutions. All that needs to be in place is the desire to help one another.

Simple spiritual guidance will lead to the desired goals of these new settlements. It will be easy to recognize shades of evolution and to determine fitness for survival. Observe what a person or society fears. Retarded and infantile societies fear loss of possessions, status and cling to material advantage; whereas cultivated and advanced peoples fear loss of freedom, justice and peace.

The future belongs to the peoples of freedom, justice and peace. They will have to constantly be on the move, grouping and regrouping like the tides of the oceans, and eventually living by the ocean to observe the way water moves and flows. Yet they must also be in their C-Enter of being, in partnership with spirit. They will flourish because their common sense overpowers internal and external desires. Those who are overpowered by their desires will vanish.

In the past, navigators, explorers, sacred travelers and curious searchers founded major cultures. In the coming years, the more mobile we are and yet the less dependent on outside resources, the stronger and more capable we will become. We will need to find new energies or learn to live without the current ones.

Sailors once discovered lands faraway. We will also need to move to the sea to re-colonize a new civilization after the long wars and sudden cataclysmic events are over. The sooner we get used to being partners with nature and the oceans, the better equipped we will be for survival. Courageous men and women will have to lead the way.

A few will make it through the rough mountainous terrain. Life by the sea and on boats will eventually be more realistic, once the earth settles down again. The sea will be a source of healthy food more readily obtainable than inland. New principles of compassionate co-existence will be more easily established by sea faring small communities accustomed to tolerance and discipline, than by land-locked people unfamiliar with cross-cultural exchanges.

Barter will prosper in places we have never heard of before. New land masses will rise as old ones sink beneath the ocean. Since we cannot predict these changes, it is best to be living above 5,000 feet where you will not have to deal with the impact of rising sea levels. The trend for the sea as a place for the displaced to regroup will grow. Seagoing vessels will be redesigned to travel by solar energy.

Future networks of people on a sacred journey will expand across the globe. Communication via radio signals and an Internet type web of life will grow. We will witness parallel societies living next to each other. The old world will die out, while others prosper through seeking ancient wisdom in untried times. There is only real knowledge when there are real people.

Pluto in Capricorn by 2009 inaugurates the crack in the wall. Those who want to lead a new space time, and be on the edge of civilization when opportunities call must be able to move at an instant's notice. They must also be trained in survival techniques. The changes will be unexpected, frequent and surprising.

The regrouping of societies, which also started to take place at that time, will continue over the coming years until 2008, when a new form of intelligent network governance will be installed.

The ancient rule for success on the move is to be able to have discernment of three key variables: Knowledge of the stars and heavens above; an understanding and familiarity with the earth below; and a clear-cut perception of the mutating human condition. The men and women who possess this ability will lead the coming migrations.

Follow the magi who follow the stars and become independent of status, wealth, and attachments. What is needed to move on, beyond the inevitable and obvious, is firmness of integrity, purity of spirit, clarity of knowing, and the resolve to adapt and learn. The sooner you adapt the better.

To explore human migration you must be in balance with heaven and earth. Having the knowledge of the Cosmos above your head, seeing what is in the skies, and having respect for what is below your feet will create the magi migrator, the navigator within you.

Inter-Dimensional Schools of Light

"Self-education is, I firmly believe, the only kind of education there is."

~ Isaac Asimov

Is the global chaos we are experiencing today, fall-out from the destruction of ancient knowledge that was taught in the civilizations of Mu, Lemuria and Atlantis, in their inter-dimensional schools of Cosmic light? Western armies, while colonising the world, wiped out the ancient cultures, destroying their stories of peaceful co-existence and cooperative. These civilizations and their schools were Cosmic contact cultures. Yet, they are now an endangered species.

We hover on the edge of destroying the world's natural habitats, on the edge of destroying ourselves through wars both civil wars and national. Food and water will soon be a scarce commodity, and water wars will follow the oil wars. If a country does not have water then how will they drive their oil rich cars? The majority of people on this planet no longer care about the plants and trees; they fear the wilderness and hide from the darkness of the night skies through illuminated town and city lights. People are not robots and yet they behave as such, never looking up at the night skies.

The masses shut out and are isolated from the great schools of the stars and planets overhead, or from the vast and ancient schools of nature on this earth. People do not see and no longer hear the wisdom of the plants, trees and rocks. They do not open their hearts to the unknown, and they have lost contact with their *ancestors* and the beings of light in the world of spirit. The inter-dimensional schools of light were the sacred contact places of the Pictish, Celtic, Tibetan and Native peoples. Humans once learned greater wonders than they do now.

The initiations of Egypt, Native American or of the Toltecs, were not hierarchical forms of creating a power elite – as in secret societies today. The individuals had to be worthy, or they would only do harm to themselves and to others. The power landscapes these apprentices entered and studied were highly dangerous places for those who were not in the flow, or who might harm the natural spirit world though ignorance.

The sacred teachings were the universities of the heart and then of the mind. Things are now changing and as we return to the ways of the light, there will once more be a city of Cosmic light to realize human destiny as a light science. It will be the center of science and art, combining spiritual truth with the technology of the people of peace, and it will also be a meeting place for worlds seen and unseen.

Around the world ancient star gates are the repository of ancient knowledge. In forgotten ways *The Master of Light* evenly distributes the energies against the tide of violence and fear sweeping planet Earth. Unlike the modern Europeans who began to take hold of the world, starting from the Roman Empire at the beginning of the age of Pisces, these older cultures of the age of Aries, Taurus and all the way back to the age of Scorpio 20,000 years ago, were connected to something real yet mysterious.

In each of those ages, including the Celtic world, which was also a part of Stonehenge and other stone rings around the world, the sacred sites were portals or doors to other dimensions. The sacred places were contact points, not only used as part of human gatherings, but also other world gatherings. These were, and are, key sites for the dissemination of sacred teachings in order to keep in touch with Cosmos.

It was at each site in Egypt, Peru, Mexico, Easter Islands, Tibet, and throughout the world that key people were educated to carry out certain functions within the community, for the purpose of guidance and protection. Across the planet were points of power where the space-time dimensions crossed. These major power points or earth chakras were connected to many areas all over the planet, like magnetic lines, or veins leading to the major arteries of the beating heart. In turn, they connected to a mysterious, seemingly invisible grid around the earth, known as the *Earth-Star* grid points.

In my childhood I vaguely recall being taken to an ancient extra-terrestrial school at night while I was sleeping. Extremely advanced beings showed me the highly complex truths of sacred geometry relating to parallels between the earth's physical landscape, the Cosmos, and the human psyche. I was shown that nothing happens on this earth outside of this basic perfect cosmic order.

The consequences of pollution are a signal that the most basic rules have been broken. Effects of the bombs destroy the sacred geometry connections of the earth, even as the physical space-time and energy fabric is torn apart. The sacred geometric symbols are part of an advanced

Cosmic language, or Cosmic truth.

As a youngster, the reason why I was interested in chess and geometry was because I recognized the ancient light-colors in those teachings. The geometric forms and lines are linear descriptions of subtle energy patterns, and each one represents some type of doorway. As a child I loved to play with these patterns and build structures that have no meaning within the disorder of this present world.

The lessons of the *Schools of Light* were not easy to learn. The teachers are from a race so old we could not calculate it; and they do not look like us. The sacred geometry was also used to bring us children face to face with a deep spiritual and karmic awareness. Many children from other neighborhoods were also taken to these *schools of the night*. However, most of them could not deal with the teachings, since they required each of us to face our own selves; this is not easy – it requires total honesty and loss of *self*.

The light schools of the Cosmic night were exacting and demanding in a way impossible to describe. Each child determined their own education depending on how we dealt with the awesome realities we were being shown about ourselves. I was shown that these inter-dimensional schools of Cosmic light are millions of years old.

Our teachers were concerned with twists and turns in our light body that cause areas of shadow that later in life reflect in our behavior. We were taught how to directly alter those self-induced twists in the energy shell through consciously changing our behavior. Needless to say, I would not discuss this experience with adults. I merely observed how dysfunctional they all seemed.

I learned that our pain is the failure to feel love, and

we return again and again to the darkness of our self-created isolation, because we continue to betray each other – which is to betray ourselves. The hurt and pain we feel intensifies only because we do not walk through the portal of those teachings of love, and into the Cosmic light.

The ancient teachers or guardians of those schools knew each soul as if each of us was their own eternal child. Unlike the institution type schools on earth, the teachers did not encourage or demand anything. The whole potential for growth, learning and change had to come from within. Something else about the schools of the night demanded complete attention and the highest level of awareness: it was everywhere, in the walls, in the surroundings, in the air and inside us. It was truth. It was the living school of the great Merlin.

I was also shown how the inter-dimensional schools of light were distancing themselves from instant or direct contact to humans, because they saw that humans were about to destroy their world. That is the sadness I felt as a kid, the uselessness of the way people live. It was also the loneliness I encountered as an adult; the inability to see or meet people who had made contact with Merlin's schools of the night.

I saw over the years how the portals were closing and the education of the children on every continent was diminishing. People were becoming dysfunctional and losing their self-esteem. Then as adults they produced children who were similar to themselves in behavior, habits and attitudes.

Finally one day, in 1991, I saw that the underlying balance was gone, and humans were becoming barbaric, debased and even more dysfunctional. I saw how people could not cope with their emotions and feelings, which were overwhelming and crushing them. The inner structural order was disappearing and leaving humans to their service-to-self lives. People were no longer learning, but instead were deteriorating as they turned away from the inner light.

A dark force or disharmonic energy was attacking the many contact points to the dimensions, shutting humans away from the teachers of light and locking the masses into a materialistic non-aware focus. The mass of people were forgetting why they had feelings and why they were alive on this planet. Most people no longer remembered their origins. They did not have the schooling for knowing their feelings and thus, do not know how to have what the teachers of light call Cosmic contact. They had lost their sight.

Feelings were beginning to come together like great clouds, angry almost at being ignored and denied their own evolution. I was shown a time when the last doorway to the dimensions closed and human beings began to go completely mad. Attacked by their own feelings and emotions, a war began on earth between the abandoned feelings and shut-down humans. Chaos is the atomic fallout of such a war.

As more and more people live only for the desires of self, they forget who they are and why they are here. The death wish and desire to destroy all ancient wisdom on earth comes from fear. The Romans began to turn away from this high level of learning and self-awareness, to serve another master. They did not want to face these hard truths, they wished only to destroy this contact for all human beings throughout the world. They wished only for power and dominance over all living things. This meant that any doorway to the dimensions that taught freedom was something to be feared and destroyed by those who serve the ego.

Why are churches and schools the foremost destroyers of the true ancient knowledge in the world? Unfortunately it is not only knowledge they are destroying. They are killing our future, the hope of humanity. World dictatorship is a dead-end future for the people on this planet. If the plans of the dark forces succeed, life as we know it will cease without warning. Dictatorship will be the end of man.

It is our own fault that we have sold out to self-serving isolation and our own greed. We are born into a beautiful world. The greater Cosmos is above our heads for everyone to see. Eternity is sitting right there in front of our eyes. We see a blue sky during the day and a dark sky full of stars at night. It is not as though a covering has been drawn over us to stop us from seeing this incredible beauty. It is obvious that above our heads is a vast incomprehensible space.

Considering the evidence that surrounds us, life is extraordinary, and it seems strange that humans do not see much beyond their wants and desires. I became aware, through the teachings of the *Schools of Light*, that our sickness goes deeper than simply wanting every material thing one can get. The real sickness is about not wanting to face ourselves on a soul level.

Having done something wrong, having moved against universal laws, humans do not want to see or face the large debt they owe, which on a spiritual level perhaps looks like something ugly and dark in the mirror of the soul. The demon that haunts us is perhaps the one we created for ourselves. Looking back at us, it is our own face that we see in the mirror, and therefore we can only overcome the demon of self, for and by ourselves.

People have hurt each other, and are in complete denial about it. They harm the earth, and are in complete denial. The dropping of mega bombs are justified through lies and deception. The *Schools of Light* teach that this is going to backfire on us. Retribution already began in 2003 when Uranus moved into the sign of Pisces.

The effects of a soul group's work stretches over many thousands of years. The evolution of a soul group happens in larger time periods than just a few lifetimes or incarnations. Geometric cycles last for periods of 13,000 and 26,000 years. The whole period contains the seeds of our developments within that time.

Unlike the western materialistically focused mind, the Native peoples, Celtic and Indian minds have used this great cycle to intensify their collective dream. The cosmology of these cultures remains to this day unrivalled for their intricate explanations, and other world knowing.

Through oppression and restriction, the predator species – most recently starting with the Romans – have continued to live according to the lower elite's dark plan. Even when the ancient shamanic cultures were being crushed and enslaved to the new world order, they remained true to their own soul group ancestor's sane yet silent development. The greater part of the work of the *ancestors* has remained largely hidden within the world of spirit sciences.

The focus of the shamans and seers took the form of a life-saving plan to counter the terror of the emerging new world we now live in. The plan is to shift the focus back into the greater spirit science energy surrounding us all through Cosmos, by working and learning together and avoiding the growing shadow-war and fear.

The time ahead will not continue to support rational development in the way we know it, nor will terror maintain its dominance over the human mind. The real war is within the psyche of man, and out of this inner conflict the physical wars manifest. We are at war with ourselves, inwardly and outwardly.

The world-wide colonial, religious, and social programming is that no one is allowed to think differently. No society was allowed to develop outside of the social rules that emerged from ancient Rome. Even though poorer and less developed countries might prosper, developing unique Celtic type systems of cooperation, they must follow the rules of Rome. Why is it that over many thousands of years no one has developed an intelligent alternative to money?

An alternative exists, but perhaps it is not so easily controlled, taxed and collected into one central account. Why do governments restrict the growth of available money as a means to control people? In a world that begins to think and act differently, these rules can and will change overnight. Survivors will have to engineer an entirely new society, unconditioned and free of the chains of the recent past.

The human DNA has being tampered with by those seeking to control this world, because the DNA light strings connect our awareness to the other world dimensions of light. The teachings within each earth-cycle are embedded within our DNA, and we can easily access those teachings, through our heart awareness. The catalyst to activating this sacred knowledge lies within the soul's emanations. Information within the DNA is activated by the conscious probing of the soul. It is essential that we open up to these *self-aware* teachings *before* the new light vibration reaches the earth.

The approaching wave of light intends to collect the brightest seeds of humankind that shine on the earth like diamonds. The tool that cuts those diamonds into form is *Contact*. How we interact with the teachers of humankind will determine what the future brings. Our conscious and focused reactivation of the inner knowing opens the doorways on earth. By opening the dimensions, much needed light is drawn into the earth and into our DNA. The teachings of this cycle are braided into our genes as patterns of light. Nothing has been lost, because we carry the entire knowledge within us.

As we come to the end of this Grand Cycle, the complete experience is contracting and concentrating into itself as a *transmutation* of the ages. People are seeing many strange things, as though the ghosts of the past and future are being compressed into one tiny space. Because of this, we have a last chance to use the doorways that help us to awaken out of this tranquilized induced sleep.

The purity and intensity of a time long past is also being concentrated into this present cycle. This is why negative or opposing forces are becoming disturbed and frightened, as they come under increasingly more pressure from the advancing light. The strangest things are about to happen, for which there are no logical answers. Strange disruptions affecting large areas will happen overnight without explanation or reason. Changes will occur and no one will know why. There will be water shortages; electronic communications will be disrupted or lost, electrical systems will fail, for no apparent reason.

A mind that does not think independently will not know how to cope with these wide ranging changes. A soul group that has spent thousands of years repressing human development will reap the harvest it has sown for itself. Its own collective consciousness will not have the intelligence to meet what is ahead. The chaos that will follow is the self-regulating light-force that will destroy the world of ignorance. Only through rediscovering our true origins will the rest of us survive.

The key to understanding what I call *Contact*, is to realize that the mind is both the navigator and the vessel; seeing, understanding, acting and moving. This is the science and alchemical knowledge we will redevelop and make available to the seekers of truth; those who can penetrate the darkness and visit the places I call the *Islands of Light*.

John F. Kennedy once said: *"Those who make peaceful revolution impossible will make violent revolution inevitable."* We have now passed beyond that inherent prediction of the young slain leader.

Those who make our human cosmic progress impossible will also make their own downfall inevitable. Other-world races who live to suppress differing civilizations create their own downfall, in the same way they impose it on others. As I am part of the *soul group* who did not abandon the Earth, I asked the others why they are returning now. They told me that when tyranny develops among weaker planetary cultures and the stronger ones sit back and let it happen, there comes a day when that tyranny will spread to those who could have stopped it, but who did nothing.

For the people on earth the future is one of cooperation rather than division. The future will be planned and developed on a community level. It is clearly a mistake to allow the larger social infrastructure to dictate community development with a dominant and uncaring centralized structure of resources concerned only with perpetuating its own existence. The structures in place now are there to dehumanize society. The service to self system makes both residents and the community dependant on a central government and a central banking system; and also, victims of a central elite group that rules over basic and fundamental resources world-wide.

By withdrawing local and community responsibility for environmental issues, the door is wide open to allowing corporate abuse. Since water and air pollution is the concern of the centralized state, important environmental decisions are in the hands of a few corrupt politicians, who can be manipulated by corporate elites. Individual governments and politicians can be more easily controlled than the local communities that suffer the consequences of drinking polluted water, or breathing polluted air.

I call this system the spreading darkness, a greed born of ignorance. Out of reach of this darkness are the *Schools of Light*. The earth itself is our teacher, with all of nature communicating with us as a one to one interaction. It is perfect in its design and execution. Above is Cosmos, the university of the skies. All we have to do is listen...

Contact With Trans-Human Souls

"Statistically it's a certainty there are hugely advanced civilizations, intelligences, life forms out there. I believe they're so advanced they're even doing interstellar travel. I believe it's possible they even came here."

~ Astronaut Storey Musgrove

We are too secretive about simple facts of Cosmic life. As a transmitter-receiver, or medium, I have accepted the commitment to bring clarity to confused and fragmented information. I am concerned with and I explore where we came from, both as an astrologer who can measure and interpret the movements of planets, and also as a being who came from the stars.

I know and communicate with ancient cultures, our ancestors and with beings from other physical and dimensional spaces in Cosmos who have made contact with the earth in the past, and who continue to remain in constant contact through those of us who have been fortunate to meet and communicate with them.

The whole UFO arena is a major source of disinformation. Rather than get confused, or led in the wrong direction, it is better to use your inner spiritual radar and inner navigator to *see* the true from the false. Together with disclosure will come an even greater confusing mass of babble.

German born NASA rocket scientist Wernher von Braun stated, and I paraphrase him: The controllers have an evil agenda. First they will use the Cold War and the *Communist* threat as a fear mongering base to reign in the free people. And then when the Cold War no longer works they will use the *terrorist threat* – à la 9/11 and Muslim *Jihad* as background. When the terrorist threat no longer works, they will try to use *Planet X*, Nibiru, Marduk, the asteroid threat, and when that deadline is over, by 2012, they will play the very last and final card... the *alien threat*.

This has been well prepared with psychological operations such as the movie *War Of The Worlds* or *Independence Day*. The goal is to put a subliminal fear of the unknown into the minds of world populations. You can see this playing out, making aliens our enemies, when in fact the *Indigo-Dragons* are the human elves or *shining ones* – seeded here to survive the earth changes and lead the light race of humanity into freedom from the known. The *New World Order* technocrats are afraid we take over with fully awakened 11 strands of ringed light DNA.

General Douglas MacArthur, Oct. 8, 1955 said: *"The nations of the world will have to unite, for the next war will be an interplanetary war. The nations of the earth must someday make a common front against attack by people from other planets."*

I don't know? Is every modern act of human society focused only on war? We are at war with nature, with the elements, at war with the earth, with each other...and now Generals predict we will soon be at war with people from other planets. I hope not, as they will most probably wipe us off the face of the earth.

In my opinion, if extra-terrestrial races wanted to fry this planet with scalar-rays, they would have done it long

265

ago. There have always been underground races of beings that exist on this planet, as well as other stars and planets within our solar system. Man has documented ancient mythologies all dealing with these strange other worlds and strange non-human beings who inhabit those worlds.

Soon these other-world beings will make themselves known to people, as the deception dies and the truth of our origins become more widely known. In the years ahead you will witness first-hand what I and a few more humans on the planet have already witnessed.

These advanced beings left behind artifacts and time capsules to show us their culture and to bring to us peaceful renewable energies and advanced space-time technology, passing from their time to ours. They bring these gifts across a bridge of light that is timeless and eternal. We already have their knowledge within our DNA; codes that can only be accessed by those who will use it for the good of mankind. We just need to learn how.

When the time is right, these beings of light will also lead us to various sacred and ancient secret sites on earth containing extremely sophisticated artifacts that will help us survive during and after the earth changes that have already begun. I feel that contact with us, at this time is part of a strategy to awaken our galactic mind, so we will figure out how to survive by developing and applying our intelligent-knowing, art, culture, science, spirituality, space-time cosmology and other interrelated tools.

If a space craft arrived, carrying light tablets of cosmology from the center of the galaxy, we would be so intent on learning about the nature of these beings that we would be distracted from learning how to make the knowledge function for us. Because humans don't want to do the work for themselves, they project the solutions outside, and look for gods, saviors and outside help. The teachers of light are there, but we humans have to create the solutions for ourselves, and that begins with internalizing the light source.

These beings, extra-terrestrials, aliens, or whatever people choose to call them, have been using time-travel for much greater spans of *time* than we can even begin to imagine. These high level beings interacted with many other cultures throughout history, and some of us living on the earth today are also from different time-spaces. Scriptures, other texts and tablets called them angels or even gods, so as to hide their actual reality, which is part of the human's great illusion.

Discoveries of older artifacts are being made, and only few of those are being reported, but many more new discoveries will be made, and those will be difficult to hide. Discoveries of our true origins must now be shared with all human beings, and so I am happy to predict the finding of key evidence. Of course, we will have to break through the rule by secrecy of this sick society, and this day will dawn after 2007, but before 2012.

Those who have access to this information, at places such as the *Installation* in Siberia, or *Area 51* in the USA, all prefer the elitist feeling of keeping the knowledge hidden, regardless of the fact that they do not even understand what they were shown. It is our duty as humans to crack open this puzzle. No government on this earth can stand in the way of human beings making contact with the inherited light DNA from across space-time. We must move beyond the few who sit on the information we need for our survival.

Only those who are sick would seek to elevate themselves above their fellow human beings and hide the truth from the people of this earth. This knowledge is not only for politicians and scientists hiding in bunkers. By keeping these subjects in private rooms behind closed doors, with all the secrecy surrounding it creates an unnecessary focus on the illicit.

When I explain to people what is known and what I know about the extra-terrestrial contact and alien issues, I risk – for now – my credibility, in some people's eyes. I'd like to see their eyes the day they know what I know. The true and known reality of Cosmic or extra-terrestrial life in and around the earth is a lot more layered and inter-related than we can explore here, and yet, you will soon see all this for yourself.

Those who came from other sectors of the universe, such as Andromeda or Sirius, are of a gene pool emanating from the zero point field of awareness, and those who came from Mars were a humanoid race made of the same gene pool as we humans are here. They lived in small numbers in underground bases within Mars. Present day society has no clue as to what exists *out there* in reality.

The ancient Picts and the Celtic nations were all aware of, and interacted with different levels of reality. This knowledge was lost after the terror of Rome. It will, however, be rediscovered and re-applied. Certain groups on earth, who cooperate with the galactic network, are fully aware of these matters and are studying alternative ways to communicate with the extra-terrestrials. People are trying now to understand the greater space of reality surrounding us, as the ages of chaos blend into a more cohesive whole.

The new Indigo kids – born from 1958 onwards – are potential leaders who are aware of source intelligence. The information we receive is sometimes stored many years ahead, to lead our decisions in the direction the galactic-mind's network deems beneficial to all humanity. This includes earth changes and preparations for these transformations; but as we know, earth will not be gratuitously wasted. We can make any decision we desire, and ultimately we will have the power to influence, via our fully awakened minds, strategic decisions affecting economic stability, ecological safety, as well as transforming the overall social order.

The galactic network is a Cosmic and global affair that includes many other races and species in outer space, and so it is time we waken up to this sobering fact. From what I know, we are a bit slow on this planet, like a third world society of Cosmos, in need of galactic development in matters of spiritual and cosmological understanding.

The advanced alien races do not co-operate with corrupt government organizations, because they are extremely suspicious of them, with good reasons. They do not regard human governments as intelligent enough to even warrant their time or energy. They choose to interact with contacts of their choice, thus creating a new leadership network, and bypassing the established order, or *disorder*.

The concepts the light beings want to show us – the culture, art, technology, philosophy and spirituality, is magnified in their presence. Which is why we have to emanate this magnification ourselves, and in doing so we become one with the new cosmology.

At an atomic level, our physical bodies are made from stars. At a sub-atomic level, our minds are non-physical repositories of a galactic mind. And at a sub-sub-atomic level, our souls are non-physical repositories of the source; I term as the intelligent Cosmos that pervades this whole universe, which, as Dr. Edgar Mitchell explained to me, is self-organizing. I would also say that Cosmos cannot be quantified or known.

The galactic minds network and the federation feel the human species can defend itself because it is of the stellar galactic mind, and of the Cosmos or first source itself. If we are unsuccessful, and the hostility spreads to other parts of our galaxy, then the federation would take notice and its members would defend their sovereignty. We shall see what happens as we move to the crossing. I feel we are looking at 2020: Node-Mars-Jupiter-Saturn-Pluto-alignment in Capricorn, which is what I would call the Cosmic strategic council at its finest.

Humanity has not yet realized how special the earth is, with its tremendous bio-diversity and complex range of ecosystems. The earth's natural resources are unique and to some degree still plentiful. It is also a valuable genetic library and the equivalent of a galactic gene pool. Other species want to own this planet in order to own this rare genetic library.

There are now extremely sophisticated extra-terrestrial cultures, integrating cosmology, spirituality, technology, culture, art and science in a proper holistic manner, that are not involved with governments, mainly because of their role with the Federation. Each galaxy has a Federation or loose-knit organization that includes all sentient life forms on every planet within the galaxy. And at some point, we too will become part of this inter-galactic federation. As much as this might remind you of star wars, it is actually close to the truth. It would be the equivalent of the United Nations of the galaxy.

The *Galactic Federation* has both observer and active members. Active members are species that conduct themselves in responsible ways, as stewards of their home planet that combine technology, philosophy and culture, enabling them to communicate as a global entity with a unified agenda. Whereas *observer members* are fragmented species still struggling with one another over possessions, land, control, money, power, and limited belief systems, which prevent them from forming an intelligent world forum for cooperation and peace.

The human race on planet Earth is a fragmented species and, for now, it is simply observed by the Federation, but is not invited into its policy making and economic systems. That is, with exception of a few who are here to prepare for the earth changes.

One of the rules of advanced cosmological spirituality is free will, which is why the extra-terrestrials might not bother to intervene against the sick and dying regressive aliens on earth. The Federation doesn't intrude on a species of any kind. It is a facilitating body, or network, not a governing force with a military presence. They will observe, advise or support us with suggestions, but they will not intervene on our behalf.

As invited members to this board of federations some of the more advanced galactic minds will be assisting humankind. But not all members have the special super defense technology, and I will now tell you what that technology is. It is called *love*. I am not talking about an imagined, airy-fairy new age type of idea. I am talking about a real and very effective, but high level spiritual "technology" based on Cosmic light. Tesla and Reich were on the right track.

Herein is the challenge. I am hopeful that the readiness of the entire species isn't the determining factor, but that a subgroup within the species might acquire the *technology* and protect the earth. The so-called prophecies of a new world of peace, comes from the *future*. The one-humanity awareness that will pervade the planet after the climate and earth changes will dramatically alter the landscape of our reality. Not only the outer physical landscape, but the inner reality will change dramatically.

Some say that high level extra-terrestrials with their super-computing technologies will befriend our more sane governments and function as allies of earth. They will set about orchestrating a unified world through a council of sages. The first cooperation of nations, after all the revolutionary mayhem, would be held in 2018, or maybe 2020 when all planets line up in Capricorn, they will supersede the UN and establish a new level of world cooperation without centralized government. In some cases this will be achieved through trickery, thus bypassing the elite groups who desire to hold onto power and control.

We have the equivalent of a decade, up to 2018, to come up with some smart answers as to how we wish to live in a peaceful *water-world*. Ideally, one wants to have non-violent defensive technologies, free, renewable magnetic energy sources, in order to detect and protect key inter-galactic crossover points. We shall see what happens.

We must review by then our preconceived notions of time and space in order to comprehend what goes, even beyond time travel, as past and future is really part of the present *now*. Only when we understand, can we change the outcome. If we humans agree about what we want, and rid ourselves of the divisive governments and fighting factions – the ones that hide all this knowledge so that they can maintain individual power and control – then we can change events before the planet reaches the point of no return. After all it is our earth.

The light beings teach a specialized form of remote viewing that enables a trained operative to mentally move into space-time, observe events and even listen to conversations, yet to take place. The time traveller is invisible to the people they are observing in the time they are traveling to, so it's perfectly safe and unobtrusive. The intelligence information gathered from this is used to develop other advanced technologies. We are entering a new world where thoughts become things at an instant's notice. Global memory reconstruction is one way we can *think peaceful events* into happening, thus changing our own future.

Active time-travel technology has been used by other races interacting with the earth. Using this advanced technology they attempt to cause changes in the fabric of time without damaging the time-line of events we know as reality. In other words, they seek key intervention points that produce the desired outcome with the least disruption. This technology requires a team of operatives, to be able to physically move into the precise space and time. Many strange occurrences witnessed on the earth are often a result of intervention in this space-time reality.

Simply stop to consider that we inhabit an unimaginably vast universe, in which you have choices to cooperate and respect nature and other life forms, or you can choose to destroy life. I am sure life did not create you to destroy life, but you have that choice. I am sure source intelligence has the same power to destroy you, just as you have the power to destroy other life forms. But it does not destroy you; it leaves you to figure it all out for yourself.

In that sense Cosmic consciousness is impartial, as it allows biological life forms to express themselves as they desire. At the highest level, where that which humans call the *Creator* operates, everything has a purpose. Even aggressive species that want to dominate other species, races and even planets have their purpose. We only have to understand the purpose. It is at this point a foregone conclusion in my mind that Cosmos orchestrates nothing per se, but understands everything, in the universal frequency of life, and transmits it forever via the galactic minds. Welcome to this reality!

There are planetary minds, solar system minds, galactic minds, and an all-engineering cosmic or universal mind, in that order: planetary, solar, galactic, cosmic. It is hard for most to imagine that each galaxy has a collective consciousness or mind, a living energy field that is the sum of all of the species present within that galaxy.

The cosmic mind is structured to be aware rather than to think, and through this awareness, creates the initial blueprint for each of the galaxies related to its galactic mind or the overall vibration of consciousness. The genetic code does not only exist as part of physical plants, animals and humans.

The code is embedded in the galaxy itself, within and emanating from the magnetic fields, space and the crystal planets. Creation is really a vibrant Cosmic DNA chain, a sacred living geometry of energy shells, spiralling into and through each other, all inter-related to one another and all part of Cosmos. It is now maybe easier to understand why we are really not alone here in this Cosmos, even if we are still children in our blueprint development.

It is not too late to activate the key to our Cosmic mind, and light the inner fire of awareness. It is left to our free will to decide, and for this we need no governments or higher authorities, but instead we must act and establish, for ourselves, contact with the greater Cosmic mind.

Cosmic Age of Knowledge

"Deep in the human unconscious is a pervasive need for a logical universe that makes sense. But the real universe is always one step beyond logic."

~ Frank Herbert, Dune

The age of Pisces and of belief has ended. Now begins the age of Aquarius or the age of knowledge. Discussions of beliefs have led us only to wars and suffering. I want to engage in the quest for knowledge and urge you to do the same, on your own. The age of knowledge will be vast and all encompassing. It will last roughly 2,000 years. I don't just believe it, I know it. Belief is based on fear, and fear is based on thought. Thought is the measuring of linear time, while intelligence is based on observation, understanding, comprehension, and truth.

The gifted psychic children born with Pluto in Sagittarius (after 1995) will emerge and within a few years be known all over the world. Many have already appeared on the scene, the *Boriska boy* from Mars and the Nepal *Buddha boy* are two who received media attention in 2004 and 2005. These new beings are here among us now, somewhat otherworldly; they have come to do their own thing in their own way.

These special kids will only laugh at their parents and teachers, and ask them: *"Are you insane?"* If asked to join the madness of this world and become *slaves to the system*. Take note of what Calgacus, Chieftain of the Picts in Scotland said, as they were about to fight the invading Roman armies, to retain their freedom.

276

"Our goods and fortunes are ground down to pay tribute, our land and its harvest to supply corn, our bodies and hands to build roads through woods and swamps – all under blows and insults. Slaves, born into slavery, once sold, get their keep from their masters. But as for Britain, never a day passes but she pays and feeds her enslavers."

The old colonizers will look into the young eyes of the children of the earth, from lifetimes ago, and realize: The global karmic debt is now due. They will at some point ask the children for help when their need is greatest. These new children will have to clean house to establish truth and justice, but they already know this fact. The other-dimensional flash of light, that is to happen in the future, is the same flash of light when Atlantis ended. It is a transformation of consciousness, of all that surrounds us. This different vibration will change the setting for the beings who are here.

We do not choose the future; rather, the future chooses us. The Spiritual was once an exact science and not something to be *worshiped*. The ancient peoples knew this.

I was looking down at the lake from high on the mountain when I experienced a massive flash of light moving out from the invisible realms. In that moment I knew I had seen this non-physical light many times. It moved across the ages in slow motion replay, across many lives, in the future, in the past. I have been revisiting this vision sometimes when awake and sometimes in my dreams when sleeping.

I saw Aldebaran, Antares and Atlantis combined. Within the vision of the spreading light, I felt and saw reality as we know it change. Most people may not even know when it happens, because we will become the new

277

reality. The opposite time-loop to Atlantis is approaching us now. These are the active twin star gates of our reality.

The change will entail instant telepathy, empathy, increased psychic powers, other dimension DNA braiding and change. Will there be more light? Or will it be that we are finer, perceiving more light, and the change of colors? There will be a shift to multi-dimensional thinking, an ability to travel through dimensions, space-time wormholes. Through direct knowing, it will seem as if it always was so. This is the quantum leap into the new era. Related to this, we will discover new ways of communication. We will know and see things in advance. Perception will deepen.

The Future comes and gathers up the beings that created light through their actions or behavior; the future looks for us, the *light workers* and kind creatures. All we have to do is live impeccably, and it will come and connect. It will find and bring us into its new reality. The energy of Cosmos comes and picks us up, connects us and plants us in a garden relating to who we are now.

You never know when that intelligence is going to shift things around. I hope you realize I am talking about a new science and not a belief system. Whether or not you believe it, is irrelevant. This will happen – it is beyond belief. It is the force at work. The sooner we step into our own power, the better we get used to working with the force. A good formula of co-creation for the future is this: Human (X) plus Cosmos (Y) creates F (Future).

Ideally, X plus Y and Future are all one, but few can merge the elements to achieve this unity.

Reality (time) does not move as a linear continuum. Reality does not *move* at all; we move and reality *is*.

Spiritual alchemy is, to me, the science of Cosmos. Knowing is its own right action. We are in the now and we move from moment to moment and in doing so we build power. Unless we live in disorder and then we lose power. Every now and then Cosmos (the future) comes and changes everything according to who we are, and are to be.

This means that events and changes happen in long count pulses, or passages of space-time, over years, generations and cycles. When the future comes and searches for us, there is a bigger circle, a larger change that is dependent on us having the power. It is Cosmos working in unseen ways.

When we build cohesive inner power and it glows around us as a light shell, the future creates us out of that light. It is actual and real, whereas the *belief* game is a powerless scam. Over the next years it will become known that the so called demons are as real as you and I. They may inhabit slightly different vibration levels, but they interact with our world in real ways. This means we have the power to change how they interact with us. If humans had developed spiritually, they could deal with this. The strategy of deception is to make humans imagine they are only physical material, and the overlords are spirit.

Many contemporary religions made the entities spirit like beings, and seemingly invincible. But in many ways they are just like us, and are just as vulnerable. Which is why the whole *terror* is cloaked and hidden in secrecy. The joke is – it is probably a technology that allows them to appear so powerful, or to inhabit more than one world. The deception is extremely complex, and when faced with multiple deceptions, it is best to know which one is the real one.

The easiest way to deal with other entity-vibrations is to know beyond a doubt that they are real, like you and I. And this is the key to defeating and overcoming them. The intelligent other world beings tell us to make peace now, to stop the wars and cooperate as one humanity.

That is the one thing the elite want none of us to figure out: that they have created a mystery out of reality. They have invented a confusing array of multiple names: Jinn's, spirits, ghosts, demons, aliens and they have gone to great lengths to program humans to believe that none of these beings exist in ways that can be dealt with. Apparently they are more powerful than humans.

When you know they exist like you and I the mystery becomes reality. I know this because the sky beings have shown me that they exist; they also teach me in the old pre-Atlantis tradition to know myself and thus to know truth. In the past, societies would never have believed the nonsense we are taught today by so-called science and religion; they would have revolted.

The world of spirit cannot interfere in the physical life, but they can and do warn us to pay close attention and to watch out for the deception. This is important, because as non-physical spirits the enemies of man cannot alter or affect mankind, but as physical real beings they can. The Native Americans, Tibetans, Aborigine people and other ancient civilizations knew the truth about this, a truth that is about to backfire on the society that suppressed it.

Some alien species over many thousands of years lost the capacity to feel. They are intellectual beings without emotions. Their spirituality has been lost together with their feelings. Similar to humans today, this loss of feeling is considered to be a higher development of their species, because they have invested their energy into high levels of technology without soul.

These beings are now discovering they have reached a dead end, in the sense that on an evolutionary level they can go no further. They are dying because of their lack of emotions. Because these beings were only concerned with advancing scientifically, they ignored the spiritual aspect of their evolution. For them, the universe was a scientific playground. Their goals and value system are based on technological advancements and little else. It never occurred to them that the emotions are the DNA navigator system and that without feelings the genetic base begins to break apart. In a secret alliance with governments, alien-human hybrids are being bred, with the hope that they can recapture their emotions.

If the physical body of a species does not recreate its energetic spin then the species eventually dies out. Our harmonic spin keeps the energy seed alive. If humans make the same mistake and lose their emotions, their genetic DNA strings will also break apart and the human race won't survive. If the souls cannot complete the physical learning cycle as physical beings, where will they go? At this point the apparently advanced technology comes to an end, because no one is there to use it. So what was the point?

More highly evolved spiritual beings would ask: How is technology advanced when it is used to kill? Whether you are using a club, a bomb or a highly lethal laser, all three are about beating someone else to death. Which means in reality these are all primitive technologies; they are not in any way advanced. The extra-terrestrial military complex technology exchange may appear advanced to humans on the ground, but what is the difference between being hit by a club, a sword or a killer laser beam? Either way people die. I would have thought this is obvious.

All forms of technological violence are equally primitive and lack cultural and spiritual development. Unfortunately people on earth have helped create this mess by destroying the guardian cultures that knew what they were doing. Those cultures include the Mayans, North American Indians, Tibetans and the Celts; they still retained a knowledge and awareness that the planet and humankind had to be shielded.

When these spiritual cultures were destroyed through colonization, the planetary and psychic shields were destroyed with them. Human awareness dropped to the level it is at today – barbaric. There is no power in this universe that can stop you from waking up now. Not even an electronic-chip implant can take command of your mind or soul when you are self-aware.

Imagine for a moment the vastness of this entire universe. Then think of the entire history of the cosmos. Then consider the unseen and eternal soul. How are they going to control this vastness through implants? If you look at the way humans build things, and how often it all breaks down, then you can imagine that their chips will also not function too well, and with a large solar flare will probably be wiped-out in one flash. Compare this mighty creative force to a tiny implanted electronic-chip. How can an occult technology chip control the whole universe?

The desire for mind control pervades all aspects of society. Why is it that ordinary people are not allowed to use the space telescopes that were constructed with their money? Why are there no space-telescope observatories in our towns? I never received astronomy lessons at school. I am a self-taught astrologer. Considering that the Cosmos above our heads is so vast, is it not strange that we all but ignore it? But if we could look out and see these wondrous things, and could marvel at the force that created and still creates all this...

When new moon and Uranus combine together, they make pretty electric moments... You may then have asked yourself why you are alive and here on this earth. Maybe you ask how you can create or how you can change. There is not much else to it than to be. Once you experience that, you are aware of it, and you hold that moment... you observe it, while being it and being as one.

That is the change. That is all there is to it. Studying Cosmology taught and showed me the reality. With one look and with mere presence we effect the change. We can then be the change we want to be and change the other. That transformation is constant; it is to be alive...

I was observing a cloud and I realized she, the cloud, was observing me also. Where it gets strange is when you tell the cloud to morph, and the cloud morphs. That is intelligent cooperation. Or did the cloud I was observing put the thought inside my head? If you consider that the observer is the observed, then who is observing who in this Cosmos? Perhaps the seeing stars make our world through their ongoing observation. Which is why I like to observe the night sky. When I look at the stars, I am seeing myself.

Strategy of Saturn

"Future Earth Security depends on our willingness to cooperate."

~ St.Clair

To understand the need for strategy and real security, I tend to turn to Saturn, the task master of time. Saturn is the organizer of order into cosmic chaos; the planet takes 29 years to revolve around the Zodiac, just as the moon takes 29 days to revolve around the Earth. The 500 year alignment of Saturn and Uranus by trine was foremost in my mind as we progressed through the years 2004 to 2005.

People in the future will have to create their own *No Fear Zone* if they are to find a sane passage through the chaos ahead. I call my own vision of the future: The City of Gold.

I feel we have to create a sane alternative to the madness now engulfing us. To understand that alternative, we must study the true history of humanity and the Cosmos. Our teacher will be Saturn, the master of time, and the aspects it will form to the other planets until it reaches the position of the Southern Cross in 2013, in Scorpio Rising.

The future is about choices, and trust, for those who enter the bridge across time. To step upon this bridge of time, I am visualizing the City of Gold; it will be a small island of light in the coming chaos.

Although *New World Order* hierarchical type organizations are moving on with their plans to destroy individual free will, I am sure it will crash land as humans have another destiny ahead of them. For now, the dark force is installing its puppets throughout the world. Iran will be next. And attacks on the Internet will increase as established institutions seek to regain control of people minds. The challenge will be to create solutions that protect communications over the next years up to 2007, or there will be no further individual development of the Internet as we know it today.

Our future survival depends on cooperation. Small isolated and competing units can easily be targeted and taken out of the system. I foresee greater cooperation between those people who begin to realize that the future of man depends on innovations. Adaptation is the key.

A spontaneous intelligence arising from system wildcats will find solutions and create a new future. Which is the most important truth I was taught as young officer in Switzerland: to be able to rethink the rules and regulations and find new solutions.

We must develop our own language of peace, as we communicate new ideas. Our freedom of speech has to be preserved. We must also find ways to preserve the ancient knowledge of the Native cultures, rather than see it destroyed by those who wish to send mankind back into the dark ages.

Uranus in Pisces until 2011 forges ahead, showing spiritual telepathic and high tech methods of information exchange, some of it coming to us from the future. I have seen the structure of a new language, in dreams that will emerge soon and change the way we think. Certain information is coming to us in coded or psychic ways that cannot be understood rationally. This new way of communicating will open the doors of our perception.

Although at present most people are polarized by their different beliefs, we are a single humanity and we need each other if we are to exist and to survive the coming upheaval.

Today's history books consider that Hitler did the wrong thing moving his troops into Russia. I see it otherwise. It is not that Hitler made the wrong moves, but that the Russians made the right moves. This is what changed history. This will become evident when the dark forces lose control in an attempt to create the new Middle Eastern divisions.

Many imagine opting-out as a non-violent and constructive way to live. But we can no longer choose to think in terms of our individual groups, families, corporations, nations, etc. More than ever we have to cooperate with one another and reach out. We need to think collectively as a global network of men and women of action.

Isolation, if everyone else falls, will not exactly help us. Just because the way we have been living up until now has protected us somewhat from what is coming, it does not mean this approach will protect us in the future.

We must unite and understand each other's cultures. We must make a point of reaching out and inter-connecting. The dilemma for the earth occupants at this point in space-time is that they lack a certain brain-mind resonance that speaks directly with the heart. This heart resonance needs to be activated, if we are to survive as a race.

Other higher races, probably from outer space, have studied the damage in human beings by sending their own people to the earth in order to heal man from inside. Star seeds are among us. Their development and sense of balance in studying and understanding the human being from inside have made them teachers and healers of people on earth.

Mistakes were made on earth that effects them also, and they wish to heal the earth as part of freeing their own soul journey evolution. Earth localizes a restricted time space only for us humans. For beings visiting this planet, the same restrictions do not apply. I am told that we humans are indeed awesome beings, albeit at present totally unaware of it, but are caught and restricted in the net of negativity and self doubt.

The men and women who form the various advanced spiritual special force security services throughout the world are also part of this dynamic of the awakening mind. Within the human being there exists an ancient awareness that speaks to us on a much deeper level than the linear mind. This intelligence is the encoded source of each human at birth.

The vibratory levels of the majority of people living on earth at this time cannot cope with the emerging reality of this beyond space-time essence. It is what I sense Neptune and Uranus will show us when it aligns to certain stars, bridging time and opening new awareness levels.

It is necessary for the future evolution and survival of the human race for humanity to enter and explore the portal of its own existence. Without the inner navigator of this ancient wholeness, the human race will become extinct within a very short period of time.

It is for us to choose whether we continue as a race or become, like the dinosaurs, relics of history. Each of us is a part of the greater whole. Each of us can and does make a difference. I have said many times, the safe place is inside our DNA far away from modern civilized or inhabited areas.

From my earliest years, I was attracted to the writings of Tolkien and *The Lord of the Rings*. As I grow older, I continue to discover more depth in this amazing work, as if it is one giant encoded prophecy. "One ring to rule them all, one ring to bind them." Or so the words go. What is the ring, other than the DNA strands and planetary moves mirrored inside out?

Human awareness is but one grain of sand in a vast universe of timeless space. Through our interactions with the earth, we have formed a fragile bridge of Cosmic travel between our dreams of the now and those of the future. To cross the bridge requires the ending of doubt.

If you break the element of trust, you create the doubt and the fragile bridge collapses. Which is why the Celts said never betray the trust, or you will never get out.

In the coming years, we will have to navigate out of harm's way. The actual survival of the species will require advanced *Lord of the Rings* high magic or Cosmic knowledge. That high magic is the merging of highest levels of mysticism and science, and more yet to be discovered for which there is no name.

It is vital in this time of crossing that those who are to survive, stop thinking in terms of enemies and opponents, and free the collective mind to see and deal with what is now happening, on a factual level. Look at the problems arising and deal with them; find the solutions and act.

Analyzing and discussing the problems restricts the energy and deters the ability to find solutions. The human mind is the most powerful tool in the universe. Do not restrict it and it will become your faithful navigator.

As individuals we can defend a family, a corporation, a country, if that may be, and why not? By all means, defend Patagonia, defend France, defend Switzerland, defend Germany, defend Russia and defend the US. But as a collective force, we can defend much more; we can defend the future of our race. Of course, first we need to see the purpose for doing this, and then care enough about it to act.

As Krishnamurti said, if we do not see ourselves and each other as part of humanity, then we are the enemy of humanity. In this sense we need a new Cosmos strategy because we all are in this together.

The new world order is not the correct strategy, because its scope is limited, and also misguided. As a collective psyche or psychic collectivity, each of us retains our unique identity, but together we form a protective shield for the future. We do need to consider that it is our own future we are protecting. It is possible that each of us may be back here, born into a world yet to come. We are looking at the far reaching unraveling of our own destiny as well as the collective destiny of humankind.

If we consider that our decisions today will effect not only those who exist now, but those in the ever approaching what is to be, then we become universal men and women, time travelers of the future. We are working with a larger sculpture than the one we now see. We are like artists aware enough to step back from the canvas and see the work in its entirety, and thus, create a new space-time together.

Each of us is accountable to ourselves alone. No force will step in and change the flow of events for us; we have to do this by and for ourselves. If we relied on outside forces, it would not be our canvas but theirs. The challenges we face on a personal level, as a nation and as a world, are what give us the chance to work on the highest level and create something new. Having the awareness that we exist in the future is enough. We do not have to see what that future will bring, but the consideration alone will guide us and our strategy. The decision is ours alone.

To continue to isolate our individual forces will result in disaster when the time of trial arrives. Each of us creates the probability of a sane and healthy society together. If individual forces are fragmented, they will divide further into conflicting parties; creating chaos and anarchy. The danger exists from within, just as it exists from without. If, however, there is a uniting force at work, a system of cooperation based on the *harm none* law, then the inner fragmentation will not take place.

In a society based on cooperation each individual becomes stronger. New types of survival communities will protect unique traditions, thus helping to create strength and support. Each of us may retain our hard learned lessons and strategies, respecting and integrating the others through our shared understanding.

Saturn is a serious task master. It shows well in its aspect to Uranus, that we repeat a 500 year cycle. 1004, 1504, 2004; we are at this point now. It is a social contract the world seems to renew every 500 years, for no reason other than ignorance.

To ascend back to the Source of Creation in spiral form is to heal issues and therefore co-create balance in our life, after which our soul spirals back to what is often referred to as home, or Cosmos. Humanity, as a group of souls, is experiencing emotions as part of a programmed reality or an experiment in time and e-motion. Electromagnetic energies create the duality or polarity of positive vs. negative; trying to find balance in order to remember its journey into ascension: The way home is closer to the stars, in a morphogenetic evolution of creation.

Future Earth Security

"If you know when you have enough, you will not be disgraced. If you know when to stop, you will not be endangered."

~ Lao-Tzu

Before continuing to explore psychic recall and its evolution, I would like to summarize my predictions from the past years, as stated in my published writings, on my web site and in radio broadcasts.

The Native cultures could see ahead and into this present age, and yet their warnings and prophecies were not set in stone. They were warnings of potential futures, and most of their warnings were passed down to us so that those futures could be avoided.

All that is written in stone is what has been, often times archived so that those who would follow might come to understand what was, what are the cyclical rhythms of nature and the heavens. Prophecies are lessons both of hope and of warning. Many of my predictions from 1999 to 2003, as documented on my Website, have already been fulfilled.

Since 2002, I have advised clients to sell everything and transform it into gold and silver. Gold and silver will double and triple respectively in value, making these both safe and profitable operations economically speaking. The refuge value remains in both metals. They can also be held safely at home in small denominations of bars and coins, and they will become the accepted tool of exchange when currencies are worthless.

Previously Documented Predictions from 1999.

* The Bush 2000 Presidential victory and ballot recount. (Made in 1999)
* A world under siege (at war) by 2003.
* The 2001 Israeli Sharon election and marginalization of Arafat.
* Switzerland to join the United Nations; fulfilled in 2002.
* Re-election of Chirac and of Schroeder, weeks prior to elections.
* The exact date of the U.S. attack in Iraq (within 24 hours precise.)
* The war in Iraq to last many years – as base to invade Iran.
* Extended military presence in Iraq and the present uprising.
* Severe weather extremes in Europe and the US.
* Loss of Russian Submarine(s).

Some Future Predictions

Predictions I made in 2001:
* Severe climate changes beginning 2003 and increasing in severity.
* Bush 2004 re-election.
* Assassinations or removal of Arafat and Sharon.
* *World War III* as a global conflict over resources.
* Discovery of a new cosmology and reality model.
* New healing technology based on using sound.
* Highly gifted children begin to appear world-wide.

Predictions I made in 2002:

* World-wide economic collapse and oil crisis.
* Martial Law in the US, with curfews, reduced air traffic.
* Massive earthquakes 2004-2011: Italy, US, Mexico...
* The discovery of a new non-polluting free-energy.

Predictions I made in 2003:

* Institution of military draft in the USA.
* Total collapse of US housing market.
* Violent storms starting 2004-2005 begin hitting USA.
* Disruption of water and power supply structures world-wide.
* Value of gold will increase dramatically.
* US revolution/civil war 2011-2017.
* Anti-gravity light ships observed over towns and cities world-wide.
* New DNA discoveries show true extra-terrestrial origins of man.
* Weather control weapons impacting the earth.
* Collapse of the New World Order.

This list is a mere excerpt of the most visible predictions. Following is a practical summary of a problem-solution type predictive narrative with some advice concerning what we can do. I am summarizing as consistently as possible all my past, present, and future predictions as far as I can remember them from the hundreds of pages of my work in the past seven years, i.e., since I started to put them in written format on the Internet.

These were the things I saw happening as we approached 2003-2006. I was not happy to see many of the events play themselves out, as I knew back then that things could be done differently. As I watched the direction move towards war, I knew that the other conflicts would then play themselves out.

From the end of 2002 and through the beginning of 2003 humans had a chance to change the future from one of extreme destruction into a co-evolution with mother earth. But people would have had to have made a conscious choice to move away from the materialistic fragment model of existence and into the Cosmos model of awareness.

The people on earth did not hear the *earth-sound*, and so they did not change their ways. The line was crossed and our time had run out. The most powerful nation on earth made its moves towards war, and another line was crossed. Since then each string of events has brought us all closer to another world war and the ultimate danger of self-annihilation.

I no longer *predict* what lies ahead for us...We are living it, and what we have to do now is focus on solutions. Using our intelligence, we have to see the fact and deal with things as they are. While our brothers and sisters fight and kill each other, we have to continue to look for (and find) the solutions to these conflicts, the pollution and the energy problem. That is the alternative, and it can still happen.

Unfortunately, world events have been *hijacked* by a small elite group of individuals, who think that only they have the right to decide the destiny of mankind. As they are on the path of darkness, they seek to lead all human beings down that road. If we do not go willingly, they will use their weapons to send us down that road *against our will*. I say, we have to power to stay with the light, regardless of the mechanical manipulations of our collective time-line.

The wars being played out are not *fated* to be, they are *manufactured* to be. The events are being put onto the map of reality by a group of people who see the only way to gain control of the world is to destroy it, and create a prison for the manageable few who are left. Having sent mankind into another trauma syndrome, they think that we will be ready for another cycle of insanity with insane rulers.

I can say now, this is not to be. Other forces and other levels do exist. These levels of light can be manifest on earth by those who live the ways of the light. Living true to source reality, such human beings would create the base for advancement in human culture and society. The key to transformation is the heart signaling to the human mind, and in doing so, creating it.

Economy – takes a downward spiral in the US/UK and Western Europe until 2009/10; global recession leading to world depression; housing market to collapse. Diminished retail will follow, with large national budget deficits, and bankruptcy of Western nations. People must eat healthy food instead of dead chemically treated seeds and genetically modified – GMO grains. Greenhouses are needed; natural seeds must be preserved. Seeds and grains will be worth more than gold after 2006. Invest in them!

2008 – March 21 – Mars Pluto opposition equinox T Square with a potential of large violence beyond description. Place your savings in gold or silver coins; keep only small monthly budgets in bank accounts. Mergers of huge corporations will occur in all media, communications, energy and transportation-related industries, such as French, UK, US and Japan auto manufacturers, electrical companies and banking corporations until eventually there will be a one world type industry in the Western world opposing a non-industrial world. Germany will cooperate with Russia and China over the next decade.

Simultaneously, many investigations into the accounting practices of financial officers around the world will take place. New security and finance laws will be implemented. Many CEO's will be forced to resign when large corporations face bankruptcy in the merger and acquisition schemes that will happen globally. Banks world-wide will merge more frequently in order to hide their impending insolvency and monopolize and control the flows of liquid cash and currencies.

Advice: Create barter economies as a solution that will prove to work eventually and be independent of the system; transform your assets; convert to Euro currency, as this currency will soon lead the world markets!

Wars: Increasing in the Middle East, and Far East, with interruptions, involving Israel, and later Russia, North Korea and China. An invasion of the US might take place from inside. The peace treaty between Israel and Palestine can only go forward once Sharon and Arafat have been removed. UK will face unusual violence due to Blair's corruption. Iraq will start to effectively defend itself against civil war as the US occupation continues. Troops will remain deployed for many years as losses will mount into the thousands. Vietnam will look like a small outing compared to what is yet to come. I predicted this in detail in November 2002.

297

Politicians: It is to be anticipated that current political leaders will be assassinated, marginalized or forced to retire or resign over doubtful financial and Iraq related dealings that will be exposed. Those who stay out of the arena of corruption are targets because they tried to resist the attempts of the corporate-political elite to control the world's resources. European weather extremes might be due to *HAARP* weather attacks.

There will be more news of corruption scandals because many high profile figures were purposefully set up for them. When they do not conform to their controller's orders, you can expect more news to break and more infiltrators will leak the information as tools of blackmail and deterrence. The truth about 9/11 will emerge by 2007. Lawsuits against the governments will total in the trillions of dollars.

Insurance corporations will become insolvent when losses mount to unbearable sums within the imploding economy. As the insurers go down, the re-insurers will perish, at which point risk-taking will become minimal. It is probably too late to counter these effects, other than to stay low key and form survival communities and small economies that trade in barter systems.

Energy: More power losses to be accounted for, more systemic breakdowns, until meltdown, are to be expected. With the blackouts will later come what I call *brownouts*, i.e., rolling blackouts which produce damage to electrical appliances and computers. Keep machines unplugged when you don't use them, and use earth-grounding techniques to protect equipment.

Advice: People will have to get used to checking all equipment, phones, cars, circuits, communication tools, machinery, etc. Keep a close tab on each issue, as well as your credit cards and bank accounts, just as a pilot does before take-off when he tests his machine. You will find that the fewer things you depend on, the less checking you have to do. The less people depend on electronic equipment the better, as breakdowns and disasters will tear apart the technical world we have come to rely on.

***Severe earthquakes*:** To be expected as of September 5, 2004 and into December 2011, occurring world-wide in many more locations and growing in intensity. I see Japan as an overlooked target, with a systemic breakdown potential of large-scale magnitude. It is difficult for me to state where there is safety from quakes and volcanoes other than to advise you to stay away from all known volcanoes and fault lines. The new Madrid line is now a research focus. California is unsafe.

***Weather*:** I predicted the extremes early in 2002 and again in May 2003. Weather patterns will move even more out of control, building toward more extremes: heat waves, large surface fires engulfing forests and towns, extreme costal inundations, dwindling supplies of fresh water world-wide, severe droughts and many more floods, until the climate virtually implodes against mankind. Seek safe areas. Only remote places away from cities, mainly by the smaller inland lakes and in remotely located mountains in much higher altitudes can balance these weather extremes. However, an area has to be hundreds of miles from known military installations to be truly safe. The Southern Hemisphere on the whole looks better than the Northern Hemisphere with its many installations.

***Isolated Incidents and Trigger Events*:** It must be understood at the outset that the so-called terrorists do not exist in the way they are depicted by the media. I assume the occurrence of at least 2-3 high profile air accidents or airplane type disasters, as well as the failure or leaks of 2-3 super power stations such as nuclear reactors and submarine losses on both Western and Russian sides. I anticipate extreme unrest in China, and a revolution of the oriental world then spreading westward.

Added to this will be computer viruses, hackers and cyber-terror attacks, bio war via infected agents, system overload, medical triage, increased addictive behavior, more lethally designed drugs, inner city collapses; prison riots and underclass riots; and deployment of biological and chemical weapons.

Environmental poisoning will take place not merely from neglect and accidental pollution but from purposeful destruction of the bio-habitat; designed to engender hunger, malnutrition and starvation world-wide. Due to reverse engineering of secret artifacts, weapons are being developed that are intended to kill off at least half the planet's population.

Advice: Invest in long lasting and basic foods and shelters. Be alert also to increased medical threats and new real virus-based epidemics, global diseases, and water contamination. Exercise your own discretion in caring for your health and where possible learn the use of herbal and alternative medicine products as much as possible. Consider water reserves and install reverse osmosis or distillation systems. Use generators and try to have back-up power. Be aware that the system will try to enforce electronic-chip and other ID systems to force you to stay inside the system. Trust your instincts to decide how far you wish to cooperate.

The aim of the would-be weather and world-controllers is to destabilize the world via self-inflicted acts of terror and threats, to enforce world control. The elitist plan is to reduce populations, incarcerate the resisters and keep a lid on the badly failing system. You have to expect more fear within society, knowing that the firemen syndrome is at work: the *firemen* are the ones setting the fire; the *security forces* are inciting the crimes.

***Security and Policing*:** There will be an overhaul of all police forces world-wide to enforce their type of order. Police will be militarized. The UN will be given more power until one day it will be eventually abolished and replaced by a consensus of intelligent cooperation with human values at heart. At stake are the issues of chaos vs. control, anarchy vs. fascism, order vs. freedom, and democracy vs. mob rules. This is the enforced passage to a saner world. It will get worse before it gets better. Create your own neighborhood watch communities! Work together! Be safe!

***Pollution*:** Will be growing and out of control until we realize we must find replacements for petroleum-based technologies and toxic chemicals. The seas will become more violent, storms more frequent. Sharks are starting to congregate in places they used to avoid. Although they are used to traveling great distances – thousands of nautical miles connecting continents and operating at many different depths and in vastly different temperatures in which humans could not live – they are now starting to grow in size and increase in numbers. This is due to the Mercury pollution we feed into the seas. Sharks now threaten to move closer to harbors and shallow waters, into canals and smaller areas, as sea temperatures are now changing.

Animals in general are becoming the silent testimony to the transition of the earth, reclaiming their territory. Don't swim in murky waters, or at dawn or dusk. These are the times and places in which sharks now congregate. I have observed shark behavior, and am told land-based animals have the same behavior. Carefully observe animal behavior and consult with animal experts.

Marine life will change, as sharks, whales and dolphins may also beach themselves due to Neptune and Mercury retrogrades, or because of the US military sonar and *HAARP* frequency experiments effecting the Cetacean navigational sonar-based survival and guidance systems.

These unfortunate trends, as well as the waste of lives and resources do not have to occur. They do not have to effect you directly as they unfold. They are here to prepare you for a new way of life, thus advising you to be diligent and circumspect through this trend that will last over several years.

Science–Technology: I see enormous breakthroughs toward the development of free anti-gravity or magnetic energy devices and free communications. Vacuum, radiant or zero point field natural sources of energy will be found. Struggles will take place to free the inventors from the system and empower people to create alternatives to meet the challenges facing us in all areas of life.

Cosmology: An entirely new theory of the origin of the Cosmos and a new astronomy will be developed. Old or *new* planetary systems will be discovered and explored. Contact with inner-earth and other world extra-terrestrials together with technology advancements will take place. A natural way of life will emerge by respecting ancient shamanic truths. Beliefs will be replaced by knowledge. New colors will emerge with frequency changes in the surrounding Cosmic light energy.

The fragmentation of our knowledge will be overcome to integrate ancient holistic knowing, healing and seeing. The multi-dimensional thought process of *seeing* via *psychic recall* and new schools that teach this process will appear. Survival centers and schools for new living will emerge. New cosmic physics will be found. The direct knowledge of our origins will be revealed.

By 2020, governments by the least qualified and most unprincipled citizens will have disappeared. Illiteracy and poverty will be overcome. Monopolies and large self serving corporations will be broken apart. Synchronicities of information will be taught. Protection and healing of the planet will occur.

Look at my track record, and judge for yourself who gave the correct information since 1999. In world affairs, ignorance is not bliss, but in fact, it is evil. The age of belief has reached its end. Beliefs are based on fear. You must not have any fears, as fear is the mind killer.

Time Portal: When the natural multi-dimensional time portal opens up, some of us are shifting to a new time line – a parallel universe where we will be on an identical earth. We won't even know it has happened but something will manifest that will reduce the human population. We will move into the spiral of the vortex.

Many people are going to collapse; because they have invested in the self-serving material world and against the earth. Mass insanity will break out into the open. At first it will form like a dark cloud around the planet, which will dissolve into nothingness together with those holding its vibration. It is wake-up time; people are now being tested. I think many had a chance to do something, yet they did not act because they thought others would do it for them. They believed in their *contract*. Beings of a higher evolution are watching us. I am told this generation of our human species is like a dinosaur, just rumbling around here on this planet and facing extinction soon.

The earth can literally wipe us out, overnight. No one will know what hit them, except that they will be dark spirits with no future bodies. To avoid this, the time of cleansing will appear. Since humans have great potential, it seems that some will survive to create a new future. I do not think this bright future will begin before 2007. But I do see the harbingers of positive change; before the spring of beauty must come the winter of our discontent.

Many interesting and triggering planetary alignments will occur in the next few years, almost too many to discuss. Briefly: Uranus opposite Jupiter and Pluto square Jupiter mirror or signal a violent series of events in 2004. Maps at the time of the opposition are ominous. Jupiter mid heaven, large, exaggerated, and Mars (war, violence) conjunct Uranus, sudden, unexpected, quake trigger, with an intersection of the South Node, or an expression of negative karma or past life debts. The potential trigger to the alignment is once a month, moon conjunct Uranus. With selected maps this trigger has proved reliable for calling quake time and location.

Be calm and seek your depth; in Pisces, merge! Violence, war-like action, confrontation, and also deception are indicated; Moon opposite Jupiter: opposition in regard to religious matters and the truth-finding effort. All of this with the node intersection calling for the release of negative karma. The above was set in motion in the early part of 2003 and will encompass those times beyond 2020.

This is a massive explosive release of energy, which can show on earth as quakes, or human-made (alien reverse-technology) bombs and shadow light explosions. As I stated before, it can be jolting as well as unpredictable. It can involve the light flash; this one will always catch you with your guard down. It encompasses massive earth movements and other explosive unpredictable behavior, also involving people's rights, world-wide.

Pluto stations in 90 degree square aspects to Uranus in 2011-2017. Uranus anchors in mid 2013 the Pluto opposition by Jupiter in Cancer, which is when things really come to an unpleasant point relating to inner security. Pluto is the lord of the underworld and in my view, the catalyst of transcendence. We can call it *Phoenix Rising*, No-Eyes prophecy as narrated by Mary Summer Rain – in Shamanic words almost beyond comprehension.

Astrological cartography map locations for Uranus and Pluto energy release show the conflict on earth. Draw a line from Norway down to Germany into the Balkans. Continue the Pluto-Uranus lines down through Egypt, Sudan, Tanzania, and continuing Southeast. The Pluto line travels on a North-South line from Russia to the Turkey-Iran border, down into Iraq, Saudi Arabia and Somalia. Saudi Arabia will fall. The west coast of the US will be terminally affected, as the great Native American Indian leaders predicted one Pluto cycle or 250 years ago, when the US was founded.

The energy to be released is of Velikovskyian

proportions. This line could be showing a nine-plus earthquake, equivalent scalar technology or other energy release. Eastern China and North Korea are also highlighted on the maps for energy release. We know that North Korea will perhaps go to war; they were trained to be the wild card.

On the whole, life in the Southern Hemisphere will be safer, more pleasant and a tad more peaceful, although much of South America will sink into the oceans. Europe will not fare differently. The US will be split apart, no doubt about that. These future maps are breathtaking in their global and cosmic implications. We will soon see how accurate they were, as we are approaching one of the three potential moments between now and the end of 2012, when the magnetic grids are activated.

ᛉ

Bridging The Worlds

Ending The Age of Fear

Psychic Defense Shield

Planetary Cooperation

Gate of Destiny

In Search of Signs

The True Force Field

Orgone Energy: Technology of Peace

Envisioning An Integrated Future

Ending The Age of Fear

"A process cannot be understood by stopping it. Understanding must move with the flow of the process, must join it and flow with it."

~ Frank Herbert,
Dune (First Law of Mentat)

If you present the right information in the right way, you can get anybody to believe anything. The most important navigation shield is *discernment*. How do you know something is true? How do you discern the nature of something? How do you know if what you see and hear is dark, or whether it is safe? What are the rules, and how do you apply them?

You have appeared inside this human body and you look out at the physical world. Your awareness, or spirit, is sort of stuck to the physical vessel through the action of an unknown gluing force, or magnetism. Before you were born into this world you could probably fly around, see and interact with the unknown. You were a Cosmic being, not a *human*. So, what are you doing down here?

I think most of us would agree that flying around in free floating mass would be a lot more fun than experiencing this 3D world that is increasingly taking the form of a toxic prison-camp. Do you have a choice? Some say not and some say you do. What do you say?

From the moment you were a very young child, I doubt that many people were too interested in what you had to say. From the first days on earth you are managed and then told what to do, what to think, how to think, what not to say, what not to think and how to behave. This is called *fear*.

If you did not conform to the stream of conditioning known as your *family*, you were *punished* in various ways. The next step was the *system*, where you are told exactly what you can and cannot do. You are given a number and that number pays the system what it is due, with the added *security* of benefits and insurance. All of which could be done in a totally different way, a way that would free you rather than possess you.

If your mind is not occupied with parroting the system's education back as was shown, then it is occupied with work and later with the media circus, entertainment, problems, struggles, escapes, confusion, old age and death. But as all this is going on, what do *you* think? Where do you originate from? Why are you here?

You may or may not consider that other world beings exist, that spirit beings inhabit what is – to you – an invisible world. It may seem irrelevant that past, present and future can merge in this moment now. At least when you die you are free of the system and have no more problems. Perhaps you will stop to think of this then and for now, struggle to survive. It is called *fear*.

Human beings fear the unknown, they fear the future, and they fear the memory of the past. They have inherited fears and they project fears. Jiddu Krishnamurti, at a public talk in Adyar, India said:

"You follow a system or mould yourself after a pattern because there is fear, the fear of right and wrong which has been established according to the tradition of a system. If thought is merely functioning in the groove of a pattern

without understanding the significance of environment, there must be conscious or unconscious fear, and such thought must inevitably lead to confusion, to illusion and false action." He also said, *"Watch it and you will be able to end it."*

Our minds are so lazy that most people do not pay attention to what is happening inside themselves. Until now most people could get through life with the way things are and a lot of painful compromise, building more fear. My point being, not anymore. The reason societies all over the world are falling into ruin is exactly because things cannot go on the way they have until now.

The system is falling apart, because the psychic field that holds us and our world together is fading. It is fading because it is transforming. It is transforming, because all of life is not stationary. Get used to it, things change!

The question of other world beings, of extra-terrestrial and inner world beings or races will soon come to the fore, as will the question of the invisible world, the existence of spirits and open contact with our *ancestors*. But the question is not about what lies *out there*, the question is about the strength of your own mind to deal with it.

To make this much clearer to you, imagine you are the pilot of a space-ship travelling through space encountering the different waves and eddies of space travel. The point is you have to know how to pilot and navigate the *space-craft* through the changes that occur in deep space. And it is no different for your body. You have to learn to pilot the physical vehicle and understand the controls.

You can exchange space craft for human body, and imagine that you are on earth inside a physical bio-engineered space craft, designed and created for you by an

311

unknown and unseen creator. When you are born nothing in the society around you teaches you how to pilot that small space exploration vehicle called your body. At this point you have to learn or teach yourself.

Other life forms do exist in this universe, some are friendly and some not so friendly, and some hate us... or, at least hate this model of space exploration. Other inner world races also exist, as well as inter dimensional worlds, races, beings and life forms. But the most important thing for *you*, is not what is out there, it is what is inside you.

Mastery of the inner abilities and controls of your own small space explorer is vital if you are to understand and interact with other-world beings and the invisible world. In some ways the many stories and experiences people share of their contact experiences are there to light your curiosity, to make you think other than you do and to get you to explore the unknown. Knowing that no explorer can move around, unless he or she has mastery of the means of travel, is the key to a future without fear.

In this sense fear is the mechanism that will stop you in your tracks, unless you observe it, become aware of it and see it for what it is. The fear is also a challenge to your restrictions; if you did not feel it you would not know something is wrong.

I was once with a group of friends, walking in the mountains on a clear starry night. We had just spent hours observing the Milky Way and watching *Twinklers* shooting across the sky. It was a very clear night.

As we were all walking home on this warm summer evening, I became aware of three beings standing under ancient trees at the crossroad ahead of us. I said nothing to the others who were with me. It was a familiar feeling, hard to describe. You just feel or know they are there and then you *see* them.

What surprised me is that the other people also sensed that someone or something was standing under the trees, and everyone reacted with fear. No one wanted to walk past the trees, and so I stopped and waited as the others moved behind me. The general fear was that they could not take *the presence*.

I stood there and I could take *the presence*, as powerful as it was. Although it was so strong it almost knocked me off my feet. I could see the three powerful beings waiting for us, and I considered the meeting a sign that everything was as it should be. I was not really able to understand why the others were so afraid, but the beings explained to me that it was because most people do not have mastery of their senses or of their inner awareness.

From this encounter I knew that the awareness inside each human being is far greater than most experience, and that events or meetings with the *unknown* trigger the greater awareness, and then most people become afraid. In some ways this is being afraid of one's own shadow, because the awareness is inside us, not outside.

People are generally afraid of the vastness of their own awareness, of their own capacities − of which they know little or nothing. It is ourselves we are truly afraid of, and to overcomes that *fear* we have to get to know *ourselves*.

At this point I would like to pay tribute to the late John E. Mack, professor of psychiatry at Harvard medical school, and leading abduction researcher. He was one of the few professionals to take the challenge of abduction seriously, whether it is by extra-terrestrials or other unknown phenomena. And because he made this move he was attacked by his peers and established colleagues.

The *attack syndrome* is played out to make weaker and more easily frightened professionals tow the line and do what is expected of them. John Mack was different; he could not ignore the evidence and sweep it under the carpet of denial. Instead he applied his skills and investigated.

In an interview with Nova he said, *"When I first encountered this phenomenon, or particularly even before I had actually seen the people themselves, I had very little place in my mind to take this seriously. I, like most of us, were raised to believe that if we were going to discover other intelligence, we'd do it through radio waves or through signals or something of that kind.*

"The idea that we could be reached by some other kind of being, creature, intelligence that could actually enter our world and have physical effects as well as emotional effects, was simply not part of the world view that I had been raised in. So that I came, very reluctantly, to the conclusion that this was a true mystery."

The extensive and thought provoking work by John Mack is outlined in his book, *Passport To The Cosmos*, where he writes: *"The alien encounter experience seems almost like an outreach program from the cosmos to the spiritually impaired."* Which is exactly my point. The people who face the alien encounter experience are the first explorers of a *new world*. It is not necessarily an outward movement; it is most definitely an inward movement. Whenever I encounter the presence of another race, being, entity or persona, I am encountering a phenomenon of consciousness. We are, after all, conscious spirit beings inhabiting the physical realms.

An interview printed by *Nexus Magazine* in 2002, about space technology consultant David Adair, describes electromagnetic fusion propulsion, and crashed discs as seen from his experiences at the secret Area 51 installation in Nevada, USA. He was describing what he experienced when he got a chance to explore the internal drive of a crashed disc. He describes climbing up and into the craft, and how it felt.

He said: *"This thing could do anything. And I really wondered who in the hell built it. So as I started coming down the outside of the engine... I noticed that now, wherever I touched the engine, it was no longer reacting with the nice blue and white swirls of energy. They had changed to a reddish-orange flame-looking pattern. And as I calmed down to try and figure out what that was, it changed back to the bluish white, more tranquil-looking pattern.*

"That's when I realised that the engine is not just heat sensitive; it reacts to mental waves. It is symbiotic and will lock on to how you think and feel. This allows it to interface with you. And that means this thing was aware."
[Robert Stanley, *Electromagnetic Fusion and ET Space Technology* – Nexus Magazine, Sept. 2002]

The UFO field is both fascinating and challenging. If it does nothing else, it gets us to imagine worlds far removed from our own. But it is only a part of the ancient legacy of the mysterious man inherited when he first appeared on earth: The legacy of spirits, faeries, elves, dragons, dwarves, angels, light beings, gods and doers of *magic*.

A strange story began to surface on the Internet in 2000, called *The Lacerta Files*. The transcripts were translated into English from the original Swedish text describing the writer's interview with a small non-human female reptilian being. The texts are strange and yet interesting.

The files discuss a secret war that is happening on earth using technology that we do not understand, and that most people are not even aware of what is happening around them. In one part of the interview the *Lacerta* being says:

"Let me say, that you are nearly as far away from the understanding of the universe as you were 500 years ago. To use a term you will maybe understand: the other species came not from this universe but from another *bubble* in the foam of the *omniverse*. You would call it maybe another dimension, but this is not the right word to describe it correctly (by the way, the term dimension is generally wrong in the way you understand it).

"The fact you should remember is, that advanced species are able to *walk* between bubbles by use of – as you would call it – quantum technology and sometimes in special ways only by use of their mind (my own species had also advanced mental abilities in comparison to your species, but we are not able to do the matter-string-bubble changing without technology, but other species active on this planet are able and this looks to you like magic as it had to your ancestors)." [*The Lacerta File – Part I*, by Jimmy Bergman, and Christian Pfeiler, Editor-Translator. Shan-Lyn Forsythe also contributed in this translation]

As I said, *magic* is its own science. You just have to learn how to use it. As humans do not know how to apply this sacred science it is called magic. But really nothing unusual is happening outside the laws of the multi-universe, it is only that we do not understand those laws and so do not know how to apply them in our lives, which means, we have to learn to apply them without fear.

So who are our friends? That is the big question. *We are!* This is why my *advisers* say to me we must establish our own inner sovereignty. Take Command! Be the masters of our own destinies – our own masters! Refuse to be overwhelmed by the apparently advanced technology of the *extra-terrestrials* and build up our own power. After we have done that, we can decide what to do about our various visitors.

Where humans are concerned, we have to get together, realize we are one race, finally get organized and establish our own power bases. Then I guess it is irrelevant who is friendly and who is not. We will simply go ahead with our own silent transformation. We must forge ahead and create our lives outside of any intergalactic, subterranean or extra-terrestrial conflicts that exist.

I feel we must, as a single planetary community, take charge and bring world affairs under our own control. That's the task at hand. If we don't do this, then no one, not even a friendly alien species can help us. Pluto and Jupiter in Capricorn with Uranus in Pisces as of 2008 are the mirror effect of that zone of change, the new time and date line marker heralding evolution of humanity, minus governments, minus greed, minus secrets, minus wars, minus control, minus fear.

Psychic Defense Shield

"Attack when they are unprepared – Make your move when they do not expect it!"

~ Sun Tzu (The Art of War)

Why would you have to develop, or have a psychic defense shield? In reality this is nothing more than having a healthy balanced mind, but when I say mind I also mean the psychic as well as the psyche. The psychic is the unseen element of the human mind that functions whether we are aware of it or not. It is the super-sense that is largely ignored, because it sends signals that the rational mind feels uncomfortable with. For example, if you are asked to do something that is wrong, the psychic sends a signal to the cortex that is later blocked by thought as the person carries out wrong action.

I say the signal goes to the cortex, because the psychic can and does send signals directly to all the organs of the body. The brain is simply one part of a complex body-brain intelligence network. You sense the world directly through the totality of the body-brain and not only through the brain inside the skull.

Following my definition you could say that people who feel highly uncomfortable, or refuse to do things they know are wrong, are people who are also highly psychic. Most people are not aware of this, because the so-called *education system* develops thought at the expense of intelligence. To fit into the system most psychics suppress

318

their skills and suffer. To be cut off from your most acute senses is not such a good idea. But to better control human populations the parasites had to trap people in their thoughts and damage or destroy the psychic abilities.

The psychic is the core energy field within and emanating around the physical human form. It is the living electrical discharge and awareness. Directing bad thoughts or feelings of anger damages the psychic field, and so for one's own sense of health and well being it is a good idea to learn to deal with anger, rather than direct it outside.

Strengthening the psychic defense shield is more a matter of one's own Zen of being. You are the one you have to be working on, and not someone else. The more you work on your own character, and find peace within, the more you enhance the magnetic shield around the body-mind.

Of course, you have been programmed to react, to feel like a victim, to blame outside forces, to hit out. You have been programmed to do all the things that damage your body and your mind, and that make you weak. But why do you accept to live like this? Why do you not have a healthy instinct to see, know and avoid what is wrong? Which means, to listen to your own psychic inner psyche.

Through an inner power you are better able to deal with the disruptive frequencies and parasitic attacks. Methods of weakening the populations include, but are not limited to all sorts of extremely low frequencies, (ELF & VLF) and electro-magnetic pulse (EMP) strikes, Scalar technology, weather, and earthquake manipulations. They can broadcast pre-scripted mind control message implants and even attack people in their dreams, by means of psychic or occult technology programming. Who are they? They are parasites of the physical and of the psyche, and they need man to live at a low negative-spin level to reach and feed off him.

This means that wars, misery, anger and fear are all good for feeding the parasitic force, but are bad for humans. We were not created to live at that level, we choose to live negatively. The inner environment is the key to self-mastery and freedom. The navigator psychic shield is the force that protects you and your journey. When people sour the inner environment they create self-made holes in their own psychic shield and open the door to dark forces. My advice is, be of good heart and stay protected and safe.

I would also recommend keeping bad dreams secret, or at least not making them public; this will block feedback from the attacking forces. Nightmares can be deprogrammed using light. Negative feelings can be sent back to the source. The brain can be cleansed through meditation.

One way to protect the mind from nightmares is to program the brain to have power when the attacks occur. This can easily be done before falling asleep by giving the command to the brain to block attacks or take control. The brain will apply the rules you give it to defeat, overcome or dissolve any nightmare attack on your psyche. This is the way of the peaceful warrior, and is not done negatively.

A more careful approach was shown to me, which consists of undoing the bad dreams. Dissolving nightmares is an art in its own right. However, it is best to use this only for disturbing or unpleasant dreams, because our dreams also contain valuable positive and desirable elements. The human brain has an incredible capacity to learn, and much learning and cleansing are achieved through dreaming.

Not all nightmares are from the sub-conscious. When you have received a particularly nasty dream, or occult type frequency attack, it is best to empower your brain straight away on waking up. Make sure any fear or horrible feeling is overcome by concentrated reestablishment of your own psychic power. You may then return to the dream mentally, and begin to dissolve the attacking elements with light frequency, or simply send those elements into the void. Only an empty bubble is left.

When being attacked, one has to take care to observe the nature of the attacks. It is as if you have to outwait it while subtly observing it. It is similar to the process of lying in wait and gathering intelligence. Even if this takes days, it is best to watch, rather than act immediately. The field of projected fear then collapses in on itself. If these forces receive no resistance, they have no power.

Always remember the mantra: What you resist will persist. Therefore, do not get caught in resistance; move around it instead! Most people react out of fear, which is why the dark force, feeding on fear, wants to use this element against humans. It is possible that the occult technology and *psychotronic* attacks cannot effect us externally unless they first activate something inside the mind, which then self-destructs from within.

From my own observations I have found that similar negative elements within the psyche respond to negativity from outside. This is why I say you should focus on the inner action, or the inner behavior. In taking command of yourself and your reactions, you are creating a strong base from which to protect yourself peacefully.

I have discovered that the light force of the psychic is the ground of the true masters. These ancestral beings are connected to the core of the human *C-Enter* of our being. If you were aware that this inner fear activates your own downfall, you would make better preparations. Emotional or psychic attacks do indeed use your own energy against you.

There are no hard and fast rules. The fact is that as you become sensitive and more aware of your own inner psychic force, you discover that you can and cannot do certain things. The things that disturb your sense of being are the things you avoid, or immediately know are dangerous. The real problem is not the outer disruptions; the problem is the inner loss of sensitivity.

When we dull ourselves down and are no longer aware or responsive that is when we do or allow things that can cause harm to the nerves, the mind or the senses. This is why the media bombards adults and children with violence. The more dense and subdued you are the less aware you will be.

Sensitivity and awareness protect, nurture and elevate. Loss of this sensitivity means that we become vulnerable to the dark side of existence, and some people become so desensitized that they become the dark side of existence. But they do it from inside, shutting down their own self-aware defense system and losing sight of the divine.

If you do not give your inner power away, then the outside force is powerless. The soul is unique and that uniqueness is somehow being manifested so that it does not just merge with the universe and become all that is.

Which means the soul is to remain its own universal and separate twin. I say this because the existence of the soul implies polarity. Something mysterious allows the soul to hold individual form inside a universe powered by a single force. Advanced scientists explain it as a feedback loop.

If the soul holds this unusual and unique individual force, it also has and is its own protection. There is a psychic immune system, allowing the soul to repel the all is one force, and thus, to maintain its uniqueness. This surely means then that nothing can enter or disturb that soul purity.

Our uniqueness is the Cosmic anti-hacker program built in at birth. Taking this further, it appears certain that the frequency attacks and mind terror induced by the psychic parasites or entities are setting off a *reaction* inside people. The weakness is not outside; it resides within, which also means that the solution lies within. There is nothing to *fear*.

If we learn, for example, not to focus outside on the attacking force, but to focus *inside*, on maintaining our own sense of balance and equipoise, then this dark force would not be able to create any feeling of mind terror, powerlessness or fear. In the moment you realize that you are at the *C-Enter* of your own incarnation and take the steering wheel of your own journey... Life changes dramatically.

The frequencies of terror are used to trigger some kind of energetic fear resonance within the unique individual. I have observed that they try to manifest attacks relating to our own *weaknesses*. Once they have set off the inner feeling of fear, we literally shake from within. But the real power of the fear we are shaking from is our own. Once the trigger is set, we take ourselves apart. Therefore we have to learn to pay attention, be aware and maintain our own peaceful sovereignty.

In taking control of my own feelings I have noticed that people who can avoid being pushed by psychic entities into that state of fear are eventually left alone, because the entities loses more energy during the attack and gains nothing in return.

The reason these psychic entities are attacking us is to feed off the terror, sorrow, fear, anger, or whatever other afflicted emotion humans carry around. Once these emotions have been activated and are being powered from within, they give off a frequency that the entities then feed from. It has been suggested that during abductions the regressive aliens also feed from negative emotions.

Low frequency *psychotronic* technology aims to destabilize the individual, or cause exhaustion through massive losses of energy when a person is continually vibrating in terror, anger or hopelessness. Therefore we should consciously try to teach ourselves to be masters of our own minds and feelings. By not allowing the emotional body or brain to be manipulated into negative states of mind, we therefore protect ourselves. I would say that the human mind is a million times more powerful than any technology, and thus will overcome any enemy.

It is up to each individual to find out how to master the mind and emotions, and not allow any outside force to manipulate them. Each thought has to be questioned, rather than acted upon. It is important to realize that the mind has the power to alter frequencies, even humanly made ones, through inner mastery and through intelligent building of orgone resonators and similar devices.

We humans are creative artists, and one of our skills is that we can build things. I dream of one day when we build beautiful giant orgone resonators to balance, protect and heal the earth. Where out places of dwelling are beautiful and peaceful earth harmonic co-existors. We can use our skills to benefit life rather than destroy it. The outer creations will also be a reflection of our inner minds.

Planetary Cooperation

"What is high will plunge to the depths and what is low will be elevated."

~ Nostradamus

To create a new society we will have to learn to cooperate in the planetary exchange of resources. It is clear that we can no longer expect to develop our own societies at the expense of the rest of the world. Just as we can cause environmental problems together with social problems, we can also create for ourselves the solutions to those problems. However, the mindset that created the problems will not invent the solutions.

The overall solution to bypassing the system is the creation and sharing of equitable trade and the barter of valuable goods, products, worthy services, skills, knowledge, and whatever is fair and compassionate. It no longer makes sense, on an economic and social level, that we continue to build monopolistic hierarchies to produce the tools we need as a society. The word *monopoly* has its roots in Greek; it means: *to sell alone.*

Ancient cultures were able to grow and survive efficiently and successfully because they planned ahead. They knew how to do this because they used what I call, *the sight*, or second-sight. These people were seers and also great planners. We have lost this ability, due to our insane urgency to monopolize and capitalize solely on technology-driven self serving resources of greed.

It is through the development of well coordinated

optimum human potential that everything, including economy and technology, exists. Everything we have in our homes is the product of ideas given to us through our imagination and vision. Then, through the hard work of a small group of pioneering individuals, our tools are perfected to become functional. This knowledge and the resulting products are shared with the wider group of society. What arises from this sharing is called development and production, or economy.

Our own inner potential as a resource is largely overlooked in a world that is divided into fragmented and isolated corporations with conflicting goals and divided agendas. The greatest potential resource of enlightened beings is the ability to share with one another, to cooperate and develop together. We must relearn this ability now.

The development of a Cosmic teaching centers is part of a larger vision encompassing the use of psychic recall, or remote viewing as a human intelligence tool to create the right choices. The future innovators of a new technology base intend to use this human resource to develop information and computer technologies on a small scale, and ultimately to share those resources with the new emerging global society.

Economy is the interaction between human beings. In understanding Economics, it is important to have a realistic perception of the future as it relates to the present. Our actions within the economy are continuously creating a specific future for ourselves. It is this oversight that a good strategist needs in order to meet and overcome any sudden changes that may appear to destabilize a society's economy.

Short term economics might answer the strategist's needs and desires today and tomorrow, but it will not meet the needs of the long term future. Mistakes made in the short term have a long term memory-effect, like an elephant. In this case the economy never forgets. It's the Mayan Long Count equivalent, as witnessed in Iraq, of social chaos, structural collapse and ongoing anarchy.

To run a good business, or a sane society for that matter, it is important to pay attention to long term needs through the manipulation or guidance of today's economic behavior. This is a concept traditional economists do not like to confront. I like to confront economists – some of my clients among them – with some hard facts:

1. Economics is the observation of human behavior.
2. There is no economy without people.
3. Economy is not a stock exchange bleep on a screen.
4. Economy is the co-operation of human resources.
5. Humanity is economy.
6. There is no money, only debt instruments.
7. Each human being is his/her own private banker.
8. Nature is the perfect functioning economy.

Perhaps we cannot meet every need on the planet, but we can make a new beginning. It is vital that we learn to understand how to deal with human interactions in a way that not only increase, but also enhance inventions, projects, wishes, hopes and needs. A true economy grows like a tree. In fact, nature is a perfect example of true economics at work, and should be studied by every economist. The earth, the sun, the planets and space itself are the perfect models for economic understanding.

To find a true balance between development and profit, the economist has to consider all aspects involved in human interaction. Money as the only consideration of the value of profit creates such an imbalance in demonstrating the function and needs of society; it has no real long term value. Today's high profits can just as easily be tomorrow's downfall.

In any economic evaluation there has to be an appraisal or consideration of social values. The social conditions in each area have to be met and dealt with as part of an overall growth of economy, together with balancing the interactions of human beings whose goal it is to survive and live in a decent way. Any gains or benefits resulting from a new project need to reflect its success within the community as a whole. Growth economics is about creating healthy people in a healthy environment. How else will we meet our needs as a planetary society?

As in nature, a growth situation benefits many creatures. The strong and healthy growth of a young oak tree will later bring food and nourishment to many creatures. It will bring shelter to animals and humans. It also manifests majestic beauty, and is important to a healthy natural economy, so it has to be included in understanding the structure of any growing financial endeavor.

Any project undertaken is like a young oak tree, that in following the rules of earth economics must stimulate many areas of growth, bringing with it balance, protection and wealth. Without social awareness, an economy becomes a self-contained tower of acquisition that in the end benefits no one.

A study of Saturn's behavior gives us a clue about understanding the relationship between social awareness and the economy. Saturn's position during the coming years indicates a transformation of economy.

When economic development benefits a greater number of people than those creating the new opportunities, society as a whole grows for the good of all. Development is merely the organization of human interaction and cooperation of human resources in a balanced way.

There is no such thing as isolated growth. When economists are aware of this truth, they are able to correct imbalances that might have the potential of creating a human disaster, when the dark-tower of isolated profits collapse. This will happen soon because economists have lost the human tool of second-sight – if they ever had it to begin with.

Because the world shamanic cultures were destroyed, people lost sight of true economics and how to make the economy work. We must understand that these cultures survived for hundreds of thousands of years. After destroying them, the isolationist *victors* have trashed our planet in less than a century. Look around you!

The Celtic-Helvetic vision of truth, integrity and honour is a simple one, the basis of which can be seen in the form of a pyramid. The most enduring resources of a pyramid are at the bottom of the structure, the base. As we get higher, the material resources become more scarce. The top depends on the bottom for support, but by the same token, the absolute pinnacle is necessary for the function of the base. No single part of the pyramid can exist without the other.

For the Celts, the uppermost part of the pyramid identifies the chieftain or wise clan leader. Before our culture was wiped off the face of the earth people were connected to the economy of nature. The original culture was based on spiritual and physical coordination of all people within the community. The chieftain was the male or female at the top of the pyramid. Look carefully at their position. They are perfectly poised in balance between heaven and earth; in contact with infinite space and yet in touch with the whole physical structure of the pyramid on the earth below.

The chieftain is in the most crucial position of all. They own nothing and yet everything is theirs. They have all resources below, and the infinite space above. Their position is that of balance, or of a seer. They have given up everything; the worldly things do not command them, and yet they are in command of all worldly resources due to their Cosmic connection.

It is not difficult to apply this to the everyday world of the economics of shared human resources. Anything this person does to increase and balance the whole, will increase and balance their own life. It is the synthesis of co-operation, which is from the Greek – *a placing together*, meaning *to place*.

Knowing how and where to place yourself is part of the sacred application of right action. The healthy model of economy is about right living, and only through right living will human beings stop polluting their world and begin to show respect, gratitude and love to the planet that sustains them. No matter how difficult things are, showing heart-felt gratitude to all that is, balances chaos.

The future of economics and of life as such, is knowing how to place something. The health of the economy is based on the health of human decisions. The education of the people is the enlightenment of the economy as a whole. It is the task of the master builder or clan leader to protect education and the individual-collective creative vision. Economy is the vision of the person sitting between heaven and earth. It is the Pyramid of the Celts.

Gate of Destiny

*"A **new creature lives in the vastness of black space.**
It mirrors its creator from afar – made in its image.
The super-natural force that brought humanity to
this point in time-space is awaiting us here now.
What are we going to do about it?"*

~ St.Clair

So-called spiritual or religious institutions all over the
world – I shall leave them unnamed – and even certain
new age and extra-terrestrial organizations claim in one
way or another that one day a very high level, super
advanced and most enlightened being will come and
change *all our lives* to something better. Or, at least the
claim is, that this being shall change the lives of the
chosen ones. Whoever they may be...

Different extra-terrestrial groups claim that different
space-beings will come and save us, pulling mankind back
from the brink of destruction. At every turn misguided
human beings look outside and dream of an *outside force*
that will solve his or her problems. I don't see it
happening. In fact, I am sure it was tried in the past and
it did not work.

Saving man from the path of his own destruction has
not stopped the repeated manifestation of the actual
problem, because we are the problem. We are also the
solution, and unless we change, our future will continue to
be miserable and bleak. Both men and woman together

have to turn from the path of self-destruction, and create this change. The responsibility is ours and ours alone.

If highly advanced ET groups were to appear on earth and clean up after us, trash all weapons and create a new society... These groups would have to stay here and baby-sit us for the rest of eternity, because we did not make the change to sovereign beings ourselves. Without someone outside telling us what to do, people would return to warfare, division, arguing and the sorrow that comes with fear. What kind of future would that be? You see now how misguided these prophecies are?

If any high level deeply enlightened being is going to come and pull you away from the brink of disaster, that being is the light and wisdom of your own soul. Listen to it – you may be surprised! The answer may reside inside *yourself*.

Everyone living on this planet knows that killing is wrong. The problem is they ignore what they know. It does not take an Einstein to figure out that killing is wrong, and yet even Einstein helped create the Atom bomb. Living violent lives we are breaking the rules of Cosmos, and we are doing it knowingly. No one from *outside* has to come tell us to stop. We can make that decision for ourselves. The brain that decided to go to war has to wake up and decide to end all conflict and manifest peace.

I know that different extra-terrestrial races will not save us because all through my childhood I consciously interacted with different ET like visitors. Some of them are from the invisible realms surrounding us. They are hidden in plain sight. I know the Nordic Blonds – not the ones who are with the Greys. I know of two groups of

Greys: one trying to help, the other doing more harm than good.

I have met the *High Elves* and also a small dwarf like race and tall humanoid beings like the teachers of Socrates, from more ancient times. I am also aware of different types of *Reptilians* – those who protect and those who destroy.

The more spiritual ET beings taught me how to deal with the invasive groups through a self created bio-feedback type mind power. I learned that the human mind is the key to the future, a future of peace and that without mastering this powerful tool of the mind – humankind will wither away into the halls of nothingness.

I learned not to get angry when dealing with lower vibrational beings. If something upsets me, I meditated a few days rather than get angry. I seclude myself and in those silent states I learn how to send out a powerful free will mind-heart signal, and I tell the energy to transform the earth vibration to peace, freedom and love.

Then I learned how to master non-linear time, and through trial and error I have found ways to use my mind to enter events from the past and heal them. One thing the Nordic Blond ET types taught me is to travel back to these situations using the power of my mind and alter them. I can travel to any time, any place and use my mind to protect myself, as well as withdrawing the psychic negative energy of those attacking me. In that sense I am my own guardian, and every human being can learn to be this for themselves.

If you consider that time is not linear, but is stacked like a vertebra and compressed, then in essence all time is now. Using the mind to protect oneself does actually work.

The star beings in the future taught me to withdraw the negative energy from any situation and replace it with the dynamic power of my own free mind. The intent to harm others returns in its own negative form. Light workers deactivate dark forces and remove negative energy lines.

It means one can use the mind to go into any situation and heal it. The event can be altered. This conscious use of the power of the mind transforms past, present and future. The possibilities are endless – restricted only by the limitations of your own mind.

The most important thing I was taught by the light beings is never to do harm, not even to my *enemies* – and not to engage in negativity which will only come back in the form it has gone out. The rule of this type of engagement is to manifest only the protection of the peaceful warrior, or smart manoeuvres of the other kind. It is part of spiritual cleanliness to disconnect and deactivate negative and parasitic energies, but not to attack, fight, or clobber them. When we use violence we become like them, dark magicians, and destroy ourselves in the process. It is a fine line to walk I have to admit.

At the end of the day, no matter what anyone else has to say, your own mind has the power and the solutions to all the problems and difficulties that may arise, and you are the artist, the alchemist, the creator of those solutions. As I once stated, we are the architect of our own inner Zen garden, aligning the crystals, plants, and decorations in our home called *mind*, and we do this via our own powerful *I-magi-nation*.

You have the power to change the world and that power is the power of your own mind. No one can take that away from you, even when they lead you to believe otherwise. No amount of mind-control manipulation can change your mind when you remain grounded in your own inner sovereignty. And no amount of trauma or bad memories can alter your future once you are aware of the non-linear time element.

No matter what hardships we face, I also discovered there is a bright diamond of light buried within each negative event. Listening carefully to my life story I learned to transform the conflict and division back into its balanced and healthy spin by uncovering the diamond of the cosmic truth. Beyond the infinite lies only your soul-mirror. Look into it and be conscious.

Once you apply the wisdom of *No Fear*, you don't have to worry too much. It works on its own and has its own dynamic orbit. As you empower yourself via your mind and begin to rework events through the power of love, you can alter the past, present and the future in the ways you see fit. This is why we are here.

In Search of Signs

"The one who says it cannot be done must not interrupt the one doing it."

~ Chinese Wisdom

Beyond the edge of nowhere, I see the future clearly. While the inmates will soon be running the insane asylum of this planet they inhabit, some sages will congregate to see what they can do to point the planet in a more appropriate direction. The Jesuits who educated me taught me to question authority. They were simply wise men, not the groups who today receive world attention in a negative sense.

The Zen masters who perfected my spiritual development suggested I should also tell you to question reality. Jesuit friends were more scientific, asking me to doubt and to double check, whereas the Buddhists were alluding to a *quietistic mysticism*, with the idea of getting by and living in the now.

I have always been interested to listen to people without forming opinions or judging them. My interest was always in search of the mysterious and the impossible with little regard for organizations, rituals or methods... None of which answered the questions I had in my *search for signs*.

At the end of a long road I discovered what I had always known: That Cosmos direct has the answers I am seeking, and that I am a star child of Cosmos... Meaning that the answers are inside.

Both the Jesuits and the Buddhists mean well, and contribute positive thoughts to a world under siege. Yet, we are going to have to take a more holistic integrated approach to knowing and existence, to awareness and intention, in order to reach something resembling cosmic consciousness, which is what the new millennium is all about. By holistic-integrated, I mean that we must think in magical terms, combining mysticism and science into a cosmic faith. We must look further and farther. We must look to the stars.

As humans we must persist, yet not resist. Because, that which we resist – change – will nonetheless persist. As we all know deep down, everything must change, will change and does change. The end of organized society, or of structures as we knew them, will also mean the end of concentrations of institutionalized power, hence of governments, nations, belief systems of the past and the end of religion as we know it.

What none of the great thinkers of the past thousand years could actually envision is perfectly possible: A free world with real liberty, proper freedom of expression, and true equality. This system could be manageable by virtue of itself, with the right to pursue happiness, at least to follow one's own destiny with a sense of responsibility.

This would be the ideal vision, the fulfilment of maximum human potential, which entails cosmic co-creation instead of the old cause-and-effect thinking, taught in schools. The gap between current reality and potential vision is one of illusion and frustration. The art of change involves facing reality, even facing down reality.

It is in the real world that surrounds us that I have found the signs leading to a better future, as well as an integrated way of living in the now. Rather than doing nothing; it is our sacred responsibility to try to change what we do not like about our lives and world. The art of change is then to replace the malfunctioning system with our own new and clearer visions, which is the first step of making our dreams of improvement into a new world. This process is close in spirit to the intent of high magic. It is an art found only within ourselves.

Faith in cosmic outcomes, the self-regulating force of the universe, can give us the conviction and instil in us the certainty that we can actually fulfill our own personal, as well as our collective and cosmic, destinies and thus be achieving optimum human potential. Before we reach this point in human evolution, we must want to manifest change. The mastery of change is humankind's challenge. However, it will also be humanity's greatest asset, for the mastery of change is vital to co-creating our evolution as human beings.

Science will eventually be able to stop the aging process, so we will live much longer lives. I am sure *The Master of Light* is more than a thousand years old, and yet he looks to be thirty, thirty-five when I meet him. Life on this planet will change by itself, and it is up to us to decide if we change with it or at least not interrupt the ones doing it.

Despite the fact that people are often intimidated by my seeing and detachment from the state of the world, they seek my opinion on how they could best implement change in their own lives. What I suggest is that we all question what we were taught and fuse our notions of science and mysticism to redefine a new space-time worth living in.

To begin with let us question the present concept of evolution. Long ago, beings of a much older race seeded the earth with *star-gold*. No one knows for sure, not even human beings know what it is inside of us that actually make us tick. I feel it must be the DNA that scientists call junk-DNA, simply because they have not yet understood what it's for. Only the ancestors have understood that Cosmos acts through us.

The creator embodied Comic star-essence into our genes and into our minds, which is a self-organizing awareness or consciousness. This star-essence acts in the same way all self-coordinating life forms act in the universe. What moves the stars and planets in perfect symmetry? The same force that created man.

Alien races are coming here, from the past and future, to find the genetic star-gold within man, because they realized the ancient ones planted and hid their essence inside the human race. It is something of a Cosmic star gold rush, only they cannot find the gold. As I said, it is hidden and can only be activated from within. This means no one can find it until humanity activates this mystery from within and for themselves.

The watchers mentioned in the *Book of Enoch* are probably what we call extra-terrestrials or aliens. But they do not openly interact with humans, as they once did thousands of years in the past. The truth is: If this inner discovery, the awakening of the DNA or the inner Celtic dragon-gold does not take place, then the star gold will return to the mystery, humanity will die and a new seed will be created. The human experiment will end. So we had better start protecting our own seed, our soul DNA, through activating its intelligence by living intelligently.

The visible aspect of this star gold is our behavior and the invisible aspect is the DNA. Planetary alignments are configured in each birth chart to create the alchemy of the soul. Each planetary alignment has a higher and lower vibration. Our earth is suffering because the ancient races, those who were teaching how to balance the higher and lower vibrations, were wiped out by someone who does not want us to learn about our true nature.

The full range of human emotions, from fear to love and all the intermediaries between, is part of an inner soul guidance system, a cosmic yet internal *wholeness navigator*. Without emotions we cannot navigate the inner or outer space, and we certainly will not be able to travel the universe as is intended for us. The whole point of trying to go out into space is to go back to our source, to find our way home, to make contact.

The sacred-geometry is simple. The higher aspects of the planetary alignments energize the radiant body of the soul, the tool through which we may journey out into the greater Cosmos. The lower aspects of the planetary alignments drain our life energy, keeping us bound to a limited sphere. This pretty little mud ball covered in sea salt we call planet Earth, is after all just that, a stop-over on a long journey home.

This earth-school is creation's safety factor, because it does not want deranged star gold wandering the universe and destroying it. Extra-terrestrial races older than mankind found a way to bypass the important teachings of inner navigation and command of universal forces, by destroying their emotions and concentrating on the rational and scientific developments. They foolishly removed the genetic wiring that transmitted the knowing of the emotional body into time-space, and they developed only their technological abilities and skills at the expense of the *emotional body*.

These extra-terrestrials learned how to space travel in limited ways, using the technology we call Unidentified Flying Objects, or in ancient terminology, *Chariots of Fire*. The fact is, there is no way to cheat Cosmos and bypass the intelligent safety net, and so their behavior has resulted in the terminal deterioration of their genetic DNA, the *Divine Navigator Alchemy*, or human life force carrier.

The extra-terrestrials, in leaving their world without having resolved the inner teachings, started the process of their own deterioration, one that is now leading the way to their extinction. They did it to themselves, and this is something we humans have to pay close attention to. Feelings are not our enemies; they are our essential inner navigation system.

Each feeling we have is part of what my tall extra-terrestrial friends call *star-gold*. Our true feelings and emotions are a powerful and ongoing education coming from our *inner navigator*, who teaches us how to balance our lives. Rather than suppress or run away from our feelings, we need to pay close attention and listen.

One of the reasons we don't always remember everything in detail, about our little trips aboard their flying schools of light, is because they realize that when they drop us back here on planet Earth, we feel unhappy. So they let us handle what they think we can deal with. There is something about these higher light beings which humans lack. The human memory is only slowly re-awakening to the Cosmic-intelligence space lesson.

The crop circles are meant to reawaken those of us who have not been aboard their crafts. One of the symbols left some time ago was the Mayan symbol for spirit. It led me to see more clearly what scientist and archaeologist, Michael Cremo stated in his books, *Forbidden Archaeology* and later in, *Human Devolution*. These books were the equivalent of a major tsunami into the *science of spirit* community. I recommend you read both, if you haven't already.

His book, *Forbidden Archaeology* documents a massive amount of evidence showing that humans have existed on earth for hundreds of millions of years. It is still less time than some extra-terrestrials have been around, but such anomalous evidence, contradicting the ridiculous Darwinian monkey-evolution theory. This has now catalyzed a global inquiry into our true origins. If we did not evolve from the apes, then where did we come from?

The crop circles are definitely the miracle of our time. This is a light language no one can explain. We are being educated in mysterious ways, using the forces of nature as our guide. The strange force uses the open fields of our existence to signal to us our own mystery. They are perfectly aware of this Cosmic truth: Their safety is our safety, their well-being is our well-being.

The True Force Field

"The nail of time dies – to live on in death – like a tree trunk."

~ Nostradamus (Prophecy for 2056)

One evening, while the lights blinked across the lake in the deep vastness of space, I asked the *Master of The Light* how he commands the force field that shapes Cosmos. The question was leading, but it was an awesome question with beauty inherent that had lingered in my mind since I was a small boy interested in magic. I knew I can command the reality I create on my own. But why was that so? How would one tap into the force field? He explained it to me in this way:

An extremely elaborate scam has been engineered over many thousands of years. The scam is transparent, but most people on earth don't perceive it as it is, they don't even consider penetrating its fragile veneer. Belief is the key to keeping people enslaved to a lie, which is why I say we must replace believing with direct knowing.

What I mean about *knowing* comes from my own interactions with this force, or intelligence that exists in the space around us and moves throughout Cosmos. This direct knowing or intelligence is a living holographic navigation system. One our awareness can navigate in the same way birds can navigate the globe.

For us humans this direct knowing would alter our approach to architecture, the way we build and design, of natural location ergonomics. Natural design is the application of scientific information concerning humans to the design of objects, systems and environment for human use. Makes sense, when you think about it.

Human populations use their immense skills in limited ways. We have the ability to calculate the movements of the stars and planets, and yet most of our species live in extreme poverty. We can design and engineer complex water systems, and yet most of our species do not have clean fresh water supplies. We know what pollution means and the effect it has on the human body, and yet we pollute our water supplies on a daily basis. What is missing to cause this massive error of judgement?

This is what I mean when I talk about the *true force field* – as it is a form of intelligent protection as well as a navigational frequency for our intelligence, as shown by and through our awareness. Our ancestors would never have accepted to drink the *water* that we pipe into our homes every day. Water devoid of life, water that has lost its oxygen and that is increasingly polluted.

If you take time to read, Jiddu Krishnamurti's "Freedom from the Known", he says: *"When you separate yourself by belief, by nationality, by tradition, it breeds violence. So a man who is seeking to understand violence does not belong to any country, to any religion, to any political party or partial system; he is concerned with the total understanding of mankind."*

Basically, our sensitivity is limited to our own sphere of identity, belief and national tradition, rather than being sensitive to and looking at the greater question of mankind. We can look at the stars and wonder, and yet we do not have a planetary awareness, which means we are not *aware*.

If you reflect on this for a moment, an immense being created *all that is* or the *what is*, and not only did it create, but it is still in the process of creating, so that moment by moment something is holding the whole universe together as it continues to expand and change still further, and yet in essence it remains the same. What I call the changeless change.

If you consider this immensity and then think that such a vast intelligence would want you to *believe* in its existence, you realize the lie. As the creator is all that is, and is inside all beings in the Cosmos, I doubt this force needs you to *believe*; otherwise we are all in serious trouble. What is required is action, living close to the divine, and that is something people do not do as they continue to destroy this world – the work of the creator.

Man physically destroys what was created by the source of all that is, and that is what I mean about the limitations and falsehoods of belief. It allows humans to destroy their natural habitat and to live outside of the divine while destroying life.

This intelligence has created free will so that beings can either walk away from compassion of the heart, or they can look inside and discover the truth for themselves. Free will means the drive to change comes from within, and this learning of the inner alignment of the divine is what creates pure s*oul gold* on its journey home.

The flimsy illusion of this global scam, many thousands of years old, has been created originally by beings who rely on belief as a framework for control. It is the earliest form of mass *mind control*. It is their manipulation of belief that helps them keep people away from piercing the thin walls of the projected lies and patterns of deceit. Deception is only as good as the people who buy into it, believe it and live its ways. No matter how good the deception, you have to get people to believe it of their own *free will*, to get it to work. I am here to show you how to end the scam.

Remember, the deception is only as enduring as the will of the people to believe in it, and who therefore deny themselves the right to tear a big hole in the cheap gauze and walk right through. Fear is the controlling factor in this belief syndrome. Who among us truly believes in God? I would imagine such a belief would cause people to end all deception.

It seems humans like to believe in inequality, rich and poor, higher and lower, and the colonizers have, until recently, continued to brutally kill entire societies and those cultures that refused to bow to their beliefs and their lies. Stay alive! Believe in the lie!

Humans have a choice, either to fall apart into fear or walk beyond the walls of this illusion. The beauty of free will is that it cannot be destroyed. It is a light, the light of awareness. If people would silently use this light, the deception would evaporate into nothingness.

Spiritual leaders and gurus are part of the hierarchy, maintaining dependence, part of the mind control to take us away from the realization that the answer resides within us. Just imagine in the moment your awareness begins to see into the nature of reality, a deity appears, or a master, or someone you believe to be higher than you. So that you focus on the image of the deity or their presence, and without your being aware of it the inner development is cut short and your new awareness energy gets fed to the deity, spiritual leader, guru or ascended master.

What people do not realize is that psychological dependence on any life form is a sign of a parasitic relationship. If you make yourself something more than others, you drain energy from those who believe in you and you become a parasite. If you believe others are your masters, you lose your energy to a parasitic entity.

There is a layered vibrational field around the planet; not another dimension, but another texture where such elusive entities exist. These entities cannot be seen by most of us, and so they easily influence the mind that is not aware.

We exploit the earth's resources and in the same way the entities exploit us. When a human appears who can see the middle world and who can interact with the other strata, the deities connect directly to this guru – through which they can receive the adoration of humankind.

The deities can and do give key people, *the psychic power to deceive,* and in return they receive the energy of worship from those who are prepared to believe that one human is more enlightened than another. It starts with the belief. All I can do is to point out the danger. It is up to you to take that first step on your own; awareness begins in your own mind.

The litmus test can be easily verified. If you are experiencing an expansion of awareness and a deity or master appears, do not focus on them or think they are more than you. Instead, ask to see what lies beyond them. A genuine light being will never claim or lead you to believe they are your master. You will perceive the power, and you will see that they are their own master, their own power. But you will not be coerced by them nor made to think they are someone to be worshipped.

Deities exist to feed your weakness, and maybe in other ways to test you to see if you are ready to take the next step: beyond. It is a self-regulating law of the jungle. People have a choice to live for the self, or move beyond service-to-self and into another realm. The deities also exist as a *psychic firewall* to filter out or block access to the higher Cosmic realms – as long as we carry with us an awareness of self, or ego. Those who move beyond the deity-master syndrome and in that moment give up self, are free to travel to other levels in which deception and desire for power do not exist.

Many near-death accounts describe worlds of light where beings are kindly and it feels good to be there. It is a peaceful place because there is no selfishness, greed, wanting or fear. Those who have had near-death experiences often find peace here on the earth after returning from the light presence, because their restless wanting has disappeared.

This experience can happen on earth, you do not have to wait until you die to be aware of something beyond death. But each person has to make it happen for themselves, through direct action of will and the perseverance of intent. To make the right decision moment by moment, to move beyond desire, fear, self importance or self-pity is the intelligent use of free will. No master, no guru, no doctrine, no dogma will ever free man, nor will it bring humanity closer to the divine. The greatest power comes from within – use it!

The most astounding discovery to be made in this 21st century – and to be proven as a scientific reality, is that our surrounding reality is indeed *psychic* and not *physical*. In other words, it is the psychic force field or that which moves unseen in the background, that makes things happen. The psychic field of nature and Cosmos surrounds us and responds to us in ways we are not fully aware of. But that is our own blindness, because life is responding to us through each moment as the now.

This truth has been hidden from those of us who have been separated, or divided from their *inner wholeness navigator*. We have been conditioned to react to what an outside force, or outside authority tells us to believe or do. Political and religious organizations have been created to twist our education, science and thoughts, to cause us to wander away from the inner truth. This is mental violence at its worst.

The true force-field generator of Cosmos is our own psychic power. This statement alone is enough to free you to explore your own psychic awareness, and to develop your own vibrational freedom. By psychic power I also mean sensitivity to life, Cosmos, nature, humankind, weather systems and the weather of the psyche. The psychic is direct awareness of the spirit inside the physical vehicle on earth, in this Galaxy.

Promoting selfishness, conflict, division, fear and wars diminishes your life force through what Wilhelm Reich called *DOR* (depleted or dead orgone radiance). Living a lie leads to a slow death. A healthy reality is formed, maintained and developed out of pure *psychic radiance*, in which the physical is the vessel or vehicle for the psyche, the psychic. The psychic is therefore the content or message itself.

For thousands of years the human mind has been educated to think that everything follows from the physical, when in reality everything follows from the non-physical psyche of the soul, earth, nature and the universe. The material body is merely a medium or means for experiencing the psychic; it is not the agent – the direct experiencer.

The development and precision of free energy devices are also related to the exploration of our psychic reality. This is also known as Tesla energy or radiant energy, as a friend once called it years ago when he explained to me the zero point field. This is indeed psychic energy, coming from the psychic background of which we are all a part – Cosmos itself.

The psychic is not merely telepathy, or mind-to-mind contact. It is something much greater. It is the force that moves the energy or ether which inventors Tesla and Reich discovered and used. The psychic is the force that creates chaos; and out of chaos it creates order. *"Out of chaos stars are born,"* Nietzsche said.

God is neither chaos nor order. The creator of all that is evolved the twin's chaos and order as universal life force. Celtic shamans know this. Civilizations destroyed by colonization lived by and with this information. They did not need to explain it in logical terms, because they were in direct contact with the daily truth that life is a psychic reality. Their minds were crystal clear. This truth isn't a matter of belief; it is pure knowledge. It is the true Cosmic science.

It is for this very reason that your mind has become a battlefield. Information, psychic and psychological warfare are part of the spiritual war the controlling elite want to hide from you. The simple truth is that the mind is free motion. They also want to use you and your energy for their own energetic survival. When you wake up to this simple truth, you will instantly change what is. You will awaken yourself from the illusion and step beyond it. You alone will realize the *psychic force-field* as a self-awakening process of Cosmic awareness.

The prison of the self, with its fears and wants exists to keep you isolated from the greater psyche of nature, earth, the planets, stars and the all-encompassing Cosmos. Illusion of self is designed to disconnect you from the natural psychic force-intelligence within you. Through your connection to this positive and all-pervasive force field, you are capable of discovering things that are so incredulous, they are beyond description. This total human capacity exists outside of thought but inside your own imagination, which is beyond thought. The psychic is a vast inner landscape intimately connected to the outer Cosmos.

I would suggest to you that imagination be read as I-Magi-Nation. The mind has to be altered, through the psychic awareness, to see wholeness as a long term goal, or destination even. The perception of the greater psyche has been turned into a spiritual high for the few, when in reality it can be part of everyone's experience. Not only is it a normal and natural part of our life force, but the mind can be taught to hold the greater Cosmic line of contact at the same time it performs mundane daily human activities. It is a sign of sanity to have deeper insight. Only this deeper mind can create an extraordinary, peaceful and lasting spiritual technology.

Orgone Energy: Technology of Peace

"And some things that should not have been forgotten were lost. History became legend, legend became myth, and for two and half thousand years the Ring passed out of all knowledge. Until when chance came, it ensnared a new bearer. The Ring came to the creature Gollum, who took it deep into the tunnels of the Misty Mountains. And there, it consumed him. The Ring brought to Gollum unnatural long life. For five hundred years it poisoned his mind. And in the gloom of Gollum's cave, it waited. Darkness crept back into the forest of the world. Rumour grew of a shadow in the East, whispers of a nameless fear, and the Ring of Power perceived. Its time had now come."

~ Galadriel, Lord of The Rings

Is it possible that *evil* has been educating humanity to hate and to fight, because it cannot use us if we are peaceful? Do the lower entities need us to be in a state of war in order to make us accessible to them? I see no other explanation. As far as I know, the original human being was not created to be a violent mass-murderer, or create wars, or begin to build weapons of mass destruction. But that is the road we have taken.

It is revealing that so many of the so-called *religious* writings are violent, or describe acts of violence, or call for acts of violence. What is spiritual about killing people? The earliest stone tables and writings contain descriptions of so-called *gods* blasting, evaporating or shooting beams of energy to eliminate large numbers of humans, or to eliminate each other. Why would a God create this magnificent human form only to have the forms kill each other? It makes no sense. In fact, it is a lie...Religions order people to kill, not God. All beings are God's children, and he/she does not have a chosen race to hold high above all others on or off this earth.

Each religion calls itself the rightful owners of the earth, the heir-archy. They allocate themselves the right to kill those who *oppose* them and their *god given* rights. They also allocate themselves the right to attack and colonize the lands they do not already own. As Zecharia Sitchin pointed out in his many books on ancient Sumer and Planet X, the 12th planet, the gods who came down to earth were extra-terrestrials who later fought each other for control of the earth's resources, and they later used man in their many wars.

These so-called spiritual writings are based on belief systems, telling people to *kill for God*. In many Sumerian accounts, entire communities are destroyed, or wiped out in an instant during some kind of struggle or *UFO* type light-weapons war. From an early age, humans are taught to kill in the name of a god, or for a belief, while being admonished to "fight for good", and to violently eliminate evil.

This makes absolutely no sense and was never intended to be so. There is no such thing as a just, or good, or holy war. Of course any war is evil. In his book, *The Wars of Gods and Men*, Sitchin narrates from a Sumerian description of a storm, or an evil wind, which travelled around in the skies. He goes on to say: "Invasion, war, killing – all those evils were well known to mankind by then; but, as the lamentation texts clearly state, this one was unique and never experienced before:

"On the Land [Sumer] fell a calamity,
one unknown to man:
One that had never been seen before,
one which could not be withstood.

"The death was not by the hand of an enemy; it was unseen death, *which roams the street, is let loose in the road; it stands beside a man – yet none can see it; when it enters a house, its appearance is unknown.* There was no defence against this *evil which has assailed the land like a ghost..."*

Although the many academic slaves to the system have tried to claim that the accounts written on ancient tables are part of an illusory mythology, they are unable to explain how these ancient accounts so accurately describe what sounds like a nuclear war. Sitchin writes:

"...an evil curse, a headache...their spirit abandoning their bodies. As they died, it was a most gruesome death:

"The people, terrified, could hardly breathe;
the Evil Wind clutched them,
does not grant them another day . . .
Mouths were drenched in blood,
The face was made pale by the Evil Wind."

Apparently, so we are to believe, there are *good* wars, those worth fighting. All forms of colonization and following genocide of the Native people have been called, *good wars.* And always the emphasis is on fighting, violence and the idea of good triumphing violently over some kind of evil. Few stop to consider that the one country in the world that considers itself the champion of freedom, is the one nation in the world that has developed *and used* weapons of mass destruction, on civilian populations.

As we move into the age of knowledge, or more correctly the age of knowing, war science focuses on even more powerful ways to turn buildings into dust, or to kill many millions of people with vaporizing energy beams. Still it is all about brutality, blasting physical life into tiny bits and using energy to smash something or burn it into nothing, whether it is a building or a human. The emphasis is on violence, human suffering and manifesting terror.

Ancient prophecies talk of the coming of a new age, a time of great spiritual development, an aeon of Cosmic progress and of peace. They talk about the use of peaceful means to establish new worlds, free of violence, filled with love and compassion. Yet how can this be true? What peaceful solutions can stand up to the brutal might of the military-complex?

The solutions exist and there are many more than you think. Spiritual awakening would appear to be powerless against guns, bombs and energy beams. What people do not know yet is that a true spiritual force creates its own peaceful, practical and non-violent solutions. A highly trained and balanced *leadership* will be required; these individuals must also go through their shamanic rite of passage to peace. They are the peaceful rainbow warriors of planet Earth.

Due to the relentless programming of violence and because of military focus on killing and destruction, people assume there is no other way to live. They place their attention on conspiracy-type dark thoughts, even though many have spoken of the approaching reality of finding

357

new ways to deal with old problems. There are peaceful solutions: Cosmic light or orgone energy technologies are more powerful than any weapon we know of on earth, and they do not kill. We can protect ourselves in non-lethal and functional ways, but only with a power from the ether of another dimension. It is a force that can be used to protect, and yet it harms no one.

Lower vibration entities will, of course, stay away from this highly evolved light force. Thus, orgone related technology is the perfect protection for those who want to move away from violence and killing. This energy would end all wars overnight. There would be no need for bombs, or for nuclear wars. There would be no desire to kill or to harm anyone.

This spiritual evolution is also a *technological* mental evolution of humanity into a magical scenario of true coexistence and learning, via barter and trading, without the need for boundaries or governments. It is ultimately beyond violence and far more advanced than people can imagine. The sadness today is that our mental focus on violence is causing us to develop highly primitive and destructive technologies – ones that are destroying us.

I have had my orgone generating cones built for me, and also I learned how to do them by myself. Mine are more artistic by now, as I like pretty living ornaments in my surroundings. Yet, be this as it may, orgone generators are to me what water is to my dolphin friends. If you wish you can build your own orgone generators for your home. In this way you are able to create your own orgone force field to generate peace and prosperity – richness of spirit.

Widespread use and development of this other dimensional light frequency, or orgone energy will emerge when humans realize they have the capacity to also create new technological means to harness the ether or magnetic energy that surrounds them. Although, in this vast Cosmos other species have developed highly advanced technologies, this does not mean they are as spiritually advanced as earth's magi. Technology is not necessarily a sign of advanced intelligence. Our true intelligence is encoded within the human DNA, as above so below.

When a species does not evolve to a high spiritual level, it uses technology to achieve its aims. This means it has to find and distribute a single power and spread that power source or fuel into making its technology move and work. There is, however, another level of evolution, and the challenge is to find this other level as our earth reality transforms around us. In my opinion, many of the scriptures are attempts to deliver this scientific knowledge, but the translations are clumsy. Nevertheless, we find in some advanced and ancient texts the correct mathematical alignments of this *technology*.

This achievement might require the same time it will take Neptune to reach deep Pisces, which is 2012 and beyond. Then other levels of evolution will make contact with the divine and what I call the *spiritual techno-enlightenment* and take advantage of a powerful technology which transfers the harmonic of light in 144,000 frequencies at 8 dimensions into power effortlessly, via focused or magical hearing-seeing-type thought.

Nikola Tesla and Wilhelm Reich attempted to make this leap, and we will rediscover this knowledge. There is no other door of perception or gate for salvation left for us. If we do not make this leap, we do not have a future. We must now tune into the light-sound of Cosmos. It is as simple as that.

Those who wish to control humanity do not wish us to make this leap. This simple orgone energy is capable of peacefully protecting us and our human development, and of dealing with any lower negative force that tries to stop us. We are experiencing violence and suffering because a hidden struggle is taking place. To end that struggle, we have to become our own masters. We must take command of the inner force rather than battling an outside one.

Even the good extra-terrestrials, who in our childhood took some of us on their faster-than-light time travels, are driven to follow a plan. As positive as it may be, it is still robotic; but they remain our friends. They are missed by those of us who had these pleasant educational and playful interactions with them. They like things foreseeable and manageable, which is why I actually liked them. They are by far more reliable and friendly than humans. It's time we caught up.

Humans have one asset that even the spiritually and scientifically most advanced beings do not possess. This is the DNA core I call the *joker-soul*. I guess we were designed for it. It is the journey of the Fool, in the Tarot. Call it *God's* sense of humor. It is this asset that makes us do wild things at times. Hand a space ship to a human and you will cause star wars of the chaotic kind. I guess this is what makes humans interesting. We are trans-dimensional trans-human souls.

A being, human or alien, that has lost its dignity will become primitive and brutal, seeking to bring others down to its level. Part of our loss of dignity is due to this lack of soul connection. It shows in the anti-magical way most people are living, fenced in by their belief and fear systems. Our water is polluted, cities smell of death, living areas are overpopulated and our food is lifeless. This is the price we pay for our co-dependency on *alien gods* and loss of command of our own inner force.

When we consciously take command of the inner force guidance, the soul navigator, nothing outside can stop us from growing and finding peaceful solutions to problems. The way through begins within. Developing free-energy is our protection, our success and our new future. Peace is also a science, as well the cornerstone of power for humans who have this free magnetic energy. The science of peace was lost when the powerful shamanic nations mysteriously left the earth.

The ancient Toltec Naguals and their apprentices knew how to transmute, to go back through the *Black Sun* to a core place of creation. The entire Toltec culture is believed to have transmuted, leaving behind the pyramids of Teotihuacan, until they were uncovered much later by the Aztecs. The Toltec civilization was dominated by the Teotihuacán culture, with its inspired ideals of peaceful behavior. These were scientists and artists who formed a society to explore and conserve the spiritual knowledge and practices of the ancient ones. The Toltecs considered science and spirit to be one and the same, since all energy was derived from the one source.

The Toltec tradition is based upon three masteries: *awareness, transformation* and *intent*. The students had to have the courage to face and know themselves, and through that knowing change their way of life. The ancient Toltec knew that our rational or linear perception of reality was a singular and isolated *point of view* that generally does not consider how we fit into an expanding, living, intelligent universe. Shifting the source of our personal power from our mind to our spirit allows us to access silent knowledge, and to create the energy necessary to transform our lives. This is the approaching star-gate awareness, beginning with the transformational 2012 alignment, and beyond.

Envisioning An Integrated Future

"Cosmologists debate the structure of the universe, its age, shape, dimensions, characteristics, and composition. They apply formula and theory, believing, or perhaps hoping, that the universe can be regarded as a thing governed by a stable set of laws. It is not. The one, domineering flaw of cosmology is that it does not account for the fact that the universe is plural, and because of its plurality, it is dynamic in ways that confound prediction and analysis."

~ Wingmakers.com

Throughout the world, things are silently and invisibly falling apart. We are approaching the critical time wave zero moment, the gateway of planetary motions from 2012-2047. Planets are now lined up to lend a techno-spiritual outlook on life in the decades ahead. This period can also be the beginning of a time of deepest introspection and inner seeing.

The currents of air and water will slowly reverse. We will simply see much out-of-control weather systems during the years ahead, which include wild-fires and devastating floods, extreme freezing, dark and red skies, the shifting of perceptions in time and space, and a new understanding of the weather mechanism as the seasons compress into a new reality.

Now is the time to look to new avenues and use our natural psychic recall abilities. Seeing the future from the future is fun as well as interesting. The words I use are not the actual substantial reality. The reality of psychic recall or remote viewing cannot be described in words. The word is a sign post, like a road sign on the motorway indicating the direction to travel in. You, the reader, moving toward, arriving at, seeing and applying the actual destination is beyond the word.

As I said earlier, we can safely assume that within a few years this world will no longer exist in its current form. Everything will have to change if we wish to have a functional life, and it will change. Therefore we too will change – the sooner the better. So, why not today?

Beyond predictions, my advice to the few, those who wish to move beyond dysfunctional behavior, is that we must do more than just survive. We must live. I pointed out also, beyond my predictions, that I have seen the future from the future, or the next present as if from far away. I can say that a few will live and reorganize a new way of life. Earth will have spiritually sane people who have regained the ancient connection of the 8th brain function, and the ability to see the *future* accurately – to see the way ahead and act through intelligence. That was my recurring vision or experience throughout my childhood.

This unusual experience defined my journey for this life. It took me outside what people describe as normal. I have already stated that I felt the world I was born into in 1959 was not at all what I would deem normal. I could feel the world was not functioning properly. Today we are seeing evidence of it, and the breakdown, as we continue along this road we are taking. However, I know we have an exciting future ahead. I experienced interesting things in this new reality. My seeing of the *vision* shows me how other worlds or other possibilities exist for us humans.

During my earliest years, the experience of having visions of the future was a normal part of my life. But I knew to keep it to myself. My parents, family, playmates, school kids, teachers and later my work colleagues did not care to hear about the things I saw. My knowledge, or knowing, was frightening to them, although amusing to me. Only my Cathar descended nanny understood my visions and *other world* meetings. She gave me good advice. She said: *"Keep it to your own self... until later."*

Well, later is now. By age twelve, the experiences and future visions increased to a high degree of intensity, taking me away from the rational world to understand and feel at ease with the world of spirit. My past lives seemed to teach me about things yet to come. This awareness made little sense to the people around me who had lost the balance and function of the 8th brain, the part of the mind that senses realities as a spirit presence here on earth. This is the part of us that connects with our heart, the divine light, or the vortex of psychic recall.

My challenge was to keep functioning in the normal world, while living within the greater invisible spirit-world wavelengths. Remote viewing, in its truest form, is inner sensing not outer sensing as people may try to understand it. You have only to imagine and understand that your mind is a remote sensing holographic mirror capable of perceiving any part of this vast Cosmic reality. To see – or view – the magnified outer reality, you look into the inner holographic mirror of your own mind. The connection is inside, not outside.

Those who become aware of this natural seeing ability

are going inside and looking from inside out. From this inner depth they can see with the eyes of spirit. The experiences I have could be called remote sensing adventures. I would call it, seeing with the eyes of the soul, or with the eyes of spirit. This is something every human being can do. It is simply a choice of how we use our energy.

Each of us has a choice either to think rationally and waste our lives in linear thought patters, or to see *what is*, the other non-linear reality and enjoy life. This seeing was a sign of sanity in ancient times. The magi of Atlantis knew how to see far.

The experience I am about to describe has since merged into my everyday awareness, when I am by myself and without interruptions. If you have read Frank Herbert's *Dune*, then you know what I am talking about. At night I would waken into an altered state of awareness, which we can call *Lucid Dreaming*. In this awakened state, remote objects such as the planet Jupiter would be magnified and enhanced beyond description, within my mind. At the same time, objects close to me would appear microcosmic, tiny and far away.

The mind is the seeing mirror of Cosmos, the mirror of the soul, as above so below, as within so without. When I say far away I mean beyond this galaxy. So, whatever I was being shown, the far away was near and the near was far away, both in and beyond space and time.

Inside each one of us is a soul mirror. This internal mirror is multi-dimensional. It is a natural function of the human brain, or mind, but it gets clouded when we think only in abstract terms. Quit thinking! This part of the brain has been blocked over the past centuries, due to the programming and wars. So, people have no recollection of who they really are, or why they are here on earth at this time.

However, aware of it or not, this inner mirror is functioning inside each of us day and night. What a gift this is, to see – and the seeing is the essence of one's being. Inner and faraway sensing is the natural ability to see or view the outer through the inner. It is about going inside to discover the mirror of the soul, reflecting eternity across time and space. The Celts, Tibetans, Mayans, Native people and Atlanteans used it. They all knew this from their Lemurian heritage long ago.

I will not even hesitate to predict that this will soon become, for many more people, a normal everyday experience. It has to be so, or we will not survive as a race. A few children born recently have recovered some of these abilities and natural skills.

A good sense of balance comes from the deepest and most silent no-thinking part of the brain. That is the part that forms first and designs the rest of the physical incarnation through each lifetime. We are the architects of our existence from lifetime to lifetime. But because we got caught in a survival based thought-matrix, we lost this inner sense of balance. The gift of creation became little more than a tool to survive, partly because certain extra-terrestrial races visiting the planet cut segments of the human DNA to more easily control us. We are a valuable source of DNA strands.

I guess this is the only explanation for the way many people move around oblivious to the multi-dimensional realities. Access to the obsidian-mirror viewing part of the brain was all but destroyed after Atlantis. For this reason most humans live in what we might call a brain damaged state, having lost their capacity to see.

When I came into my body, I soon became aware that I was being damaged by my surroundings. Yet, by secluding myself, I somehow retained the ability to build my own immunity to insanity. I also argued my points with everyone, as rationally as possible at my young age. I would ask one question: *Why?* I knew they would be unable to answer. So I told them if they could not answer my questions they must leave me alone. It worked. They left me alone. I was given the freedom I wanted, to progress at my own pace.

I then chose my books and tutors carefully. I rejected those professors the others wanted me to listen to, and basically I became a major pain in the neck to both my parents and the system. But since I was more attuned and also academically more coherent than those around me, this was the deal we struck: I could have it my way, and they would leave me alone. I knew that many natural gifts were being deliberately suppressed by society. But I kept this awareness to myself – until now. Now is the time to show how we will escape the inevitable disasters.

Make no mistake; only few will find their way out of the matrix. If you are into security and you depend on an outer system, then the matrix is your compromise security blanket – while it feeds off you. To change this each one of us has to learn what I have termed *psychic recall*. Many people have heard of remote viewing; which is a small part of psychic recall. Remote viewing was created by the military and other malevolent intelligence agency planners in Russia and the US, to hit on military and political targets. The Russians were the first to develop

RV and the Americans followed. Their combined experiments turned out to be a misuse of our spiritual and psychic skills.

With RV the operator (remote viewer) is given a target, and they use certain techniques to view, spot or influence the target. I would estimate this method uses about one percent of the capacity that is available for this type of psychic mental activity. With remote viewing, an outside target is chosen by the military or the viewer. With psychic recall it is the opposite: The *seer* is the target for a higher force.

Remote viewers use their awareness to discover something outside of themselves. Psychic recall is an awareness that connects and listens inwardly to the background information. What you need to know or see comes into focus from within. It is an intelligence operating on its own, without thought.

The first remote viewers also began to have unusual spiritual experiences and encounters during some of the sessions. They were not hallucinating; spiritual forces were showing them the consequences of what they were doing. The military was training them to harm or influence chosen targets or politicians, and to influence decisions. Some of the wiser people got out of these programs when they realized the far reaching consequences and the karma they were loading on themselves.

Human beings are diviners and have the divine gift of psychic recall built into the genetic system. People are using it all the time in limited ways, but are largely unaware of it. We are naturally and continually locating ourselves in time and space using abilities we are not aware of. This is psychic recall, but in most people it exists as a subconscious action. Psychic recall is simply making people aware that another level of awareness is there, and they can tune into it.

Remote viewing is about serving a limited agenda or system. Psychic recall is about serving a higher intelligence. We are all connected to this divine or Cosmic mind, we only have to make the decision to work with it. It is our choice. Needless to say, the gift of the seer is not always a welcome gift, and yet I am grateful that at an early age I began to learn how to heal myself and reconnect. I could not accept the way things were in our world. It simply made no sense.

Survival awareness is five percent of the inner psychic recall and Cosmic sensing gifts we are born with. We have been programmed as a race to focus on survival, and this self-survival focus has taken us away from oneness with spirit. We are not here merely to survive; this would be boring. We came here to fulfill a higher purpose.

If you consider your own inner awareness and mentally extend it beyond the boundaries set for you, you may feel an acute sense of danger, or fear. This is because the limitations built into thought are there to stop you looking beyond and perceiving the unknown. However, when you look further, you will merge with inner creative forces that are beyond survival.

How can we reach optimum human potential? We must extend our inner energy awareness beyond its normal limitations and direct the mind beyond survival, into spirit. This is how we will become Cosmic light, enabling us to go beyond earth changes and the illusion of the matrix.

We must manifest the inner no-fear zone, and only then will we create the outer world *in our image*. As above, so it is below. As within, so it is without. We are the world we want to manifest. Each one of us can build our own *City Of Gold* or contact center, and create a world of functional peace.

If you think about it, it is not possible to see the future from here. The eyes of spirit see the future from the future. This is important if one is to understand that psychic recall is the ability to bridge reality, by moving through what can be called one's own inner hyper-space.

Psychic recall is simply the ability to have seen an event from the vantage point of the future looking into the past. To see the future from the future, takes only a few seconds of deep concentration. Anyone can do it, when they align themselves with the forces of integrity, light and truth.

We have there inner radar skills so that we can better adapt and make wiser choices. Astrology is the rational measurement for my psychic recall skills, a gift many cannot understand, and one that has been here for us since the beginning of time, if ever there was a beginning of time.

Seeing is a mirror, an all seeing obsidian mirror. It is the silent observer inside all of us. It is non-movement, an absolute stillness, a state of mind that exists day and night whether we are aware of it or not. Seeing functions outside of thought. The art of shifting your own awareness is the key to knowing or entering that world. It is the 8th brain or 8th sense that takes in the information and uses it so we can observe the past, present and future as one reality.

Coming as close to the inner mirror as possible gives the 8th brain in psychic recall mode the ability to see the most distant objects, persons, realities or star systems. Limits are placed only by the mind using the tool. This is why it is important to transcend the limitations and restrictions of thoughts and beliefs.

Have you ever had the feeling the moon is watching you? This happened to me during one of the last eclipses. It is a powerful feeling. As I looked up at the moon on an eclipse evening I heard the words: *"The first outpost."* When I looked into the orb of light, I immediately felt close to my inner watcher observing myself as if from up there. Some people seem to go crazy during the eclipse times, or even during any regular full moon. But in fact, when we shift with this powerful awareness, we are in tune with the Cosmos, and in tune with the unknown.

If more of us had already integrated the 8th brain function into our daily lives, things would now be different. The rational part of the brain misbehaves because it tries to short-circuit and control the psychic recall process that is taking place in the 8th brain's deeper resource. I have found that the more I use this seeing, the more it merges into my everyday reality.

As there is a merging of the future seeing into the past, the scope of the mind increases, allowing for the grounding of many realities with style and ease. The more ease with which we deal with multiple realities, the more grounded we become. In the times ahead we need to be better grounded. We must activate and use the 8th brain. We have to see far to go far.

The brain responds creatively to multiple realities. It begins to wake up when it gets to connect in a natural way. Disconnected and self-serving humans are simply unaware, because the *inner spirit* connection is missing. This means the person looks outward rather than within, and does not understand reality. We can say that self-serving action is rational thought looking out, in fear. In those moments of selfishness, the individual must turn around and look within. When we look within we see the outer world with greater clarity.

Being aware is essential to changing our world. It is the inner and not at the outer event, through which we empower ourselves across time and space. Look inside and open the lotus flower of divine seeing. Why is the lotus flower said to have eight petals? Why eight chakras and eight wheels?

The white lotus is a symbol of Buddha-nature, purity and perfection. The eight petals correspond to the *Eightfold Path*. However, it is important to realize that all these teachings and symbols are alchemical reflections of a Cosmic order, or sacred geometry. This is a living reality, and it has become a symbol, only because humans are walking outside of the living reality of balance and purity.

We are looking in from outside, rather than *being* from inside and applying the sacred geometry of Cosmic knowing. The inner mirror of seeing opens when the individual looks within. The inner eye is the force with which we see the inner light, which is the next stage of our development.

Do not take the earth for granted, or imagine that Cosmos is a *thing*. I was watching the sun rise, very early in the morning, and as the light touched my forehead I heard, felt saw:

"The son of God is the *Sun* of God." Which came to me in words, but the actual reality of what I was seeing was not the word. I continued to see, feel, be aware of the light from and of the sun, and knew I was perceiving a living being, who was teaching me – giving me *knowledge*, making me aware... because, I was and am aware. The silence of this realization filled my new mind.

Envisioning the integrated future of mankind means to understand the changes ahead, which means to understand yourself. For this, you need no method, nor any spiritual hierarchy. *The Master of The Light* never taught me to form or use a system, in order to see. All questions such as: "How can I be more aware, how can I see clearer, how can I be more psychic?" are questions of the self, and the self cannot be aware, see clearly nor develop psychic abilities. Why? All of this comes into being only when the self is not active.

Teaching systems are delusional spiritual crutches. How do I reach enlightenment? Strangely he never mentioned nor promoted individual enlightenment. He would eventually show me the doorway to collective enlightenment. He never offered me a goal, or a final summit to attain. I did make it to the absolute top of a mountain, high above lake Geneva at 12,000 feet, and rather than enjoy some sense of achievement, I saw the four cardinal directions, and far away, only more mountains.

Seeking enlightenment for oneself is egotistical. True enlightenment is when all of mankind is enlightened. We work for the enlightenment of all. How can a non-enlightened being work for the enlightenment of all? By knowing that it is right action without thought.

The Cosmic Key

Horoscope Event Mandala

When Neptune Is In Pisces

Dawn of The Time Lord

Cosmic Star Navigators

Power of Dragon Elves

Way of The Magician

Total Change

A Future Vision

The Future Is Now

The Empty Hour Glass

Entering Enlightenment

Horoscope Event Mandala

Aries 9 – A Crystal Gazer: *"Symbol of man's capacity for bringing an entire universe within the purview of his mind. There are higher faculties of human understanding, but as no more than the illimitable possibilities of personal experience in the realms of potentiality. Implicit is the concept of the world and the individual as irrevocably in partnership – each reflecting the other and thereby providing a foreview of every possible mood or situation."*
Keyword: ACUTENESS.

Marc Edmund Jones,
The Sabian Symbols in Astrology, 1953

The solar eclipse of 29 March 2006 sees the emergence of this book. The Sabian symbol, for Sun on March 29 is seen in the above 9th degree of Aries. The eclipse of Aries 9 will be observed, in 2006 directly in Turkey, Libya, Russia, Tchad, and Nigeria; and it will be interesting to see what this means – after *Zen of Stars* is published and in your hands.

A birth, or event, is a moment that is captured by planets and stars in what amounts to a photography of those planets and stars over the time and place of the event. Usually astrologers look at this in two-dimensional ways, and the whole event, known as a birth chart, is really a schematic drawing, frozen in time, and of course it is not very accurate in the sense that we must see this

Cosmos as a multi-verse, as opposed to a uni-verse. It is best to imagine the star signs in a sphere, not a circle.

Once we start to think at least spherically, in three dimensions, we get a better picture of what we are about, when seen from outside this earth plane. The signs are in a mathematical relationship to one another, and so, the planets moving in those signs form connections which mirror down into our own essence and existence. This astrological *sign language* is the subject of my two earlier books. Michael Donovan also explains this extremely well in his beautiful book – *Letters Upon The Mast*.

To me, my form of *Star Zen* is both a way of seeing the stars and planets at our birth, and is also a sort of energy meditation that pacifies the mind and heart. The whole thing is a form of magic or permanent life path in motion – leading to spiritual enlightenment in the best of both worlds, and at least it enhances our birth sovereignty.

Star Zen teaches that certain issues in life arise from the three afflictions of life: **anger, desire and ignorance** – all born of faulty conditioning of the mind, or of some bad astrological karma as the case may be. One can say this: *Anger* is cardinal, *desire* is fixed, and *ignorance* is mutable – by nature of terminology – in astrology.

The birth chart of a human being is a psychic energy shell, or an imprint of the time quality at hand, and it is for sure a vortex force field inscribed in the DNA, in form of a planetary *mandala*, or hologram in progressive and invisible motion; one that can be used as both an object for meditation or as an ongoing inspiration when you know where the planets are at any moment in time over your chart.

This planetary hologram is also a tool by which we can

378

discover the most probable future and the causes of our specific problems and chances, with its risks and opportunities inherent in our existence here now. We are the ones who must understand our birth mandala, and work with our energy shield – one that is unique to each human being unless they are twins born in the same minute.

The horoscope energy clearly shows the destiny potential as well as the fated course of events, or the most probable future. The fixed stars, planets, signs and houses clearly indicate the mental and emotional as well as the parental conditioning that has distorted our way of perceiving life, and indicates what is causing us difficulty in attaining enlightenment and in accomplishing our mission, which we signed by coming into this realm.

For now, we shall consider the signs of the zodiac by their quality (cardinal, fixed, mutable) and how they may reveal individual dharma, or truth. The energy of the signs, as we know, can be expressed both positively and negatively. Negative expression occurs when we are acting or reacting in a conditioned manner. And who can truly say he or she is totally un-conditioned?

Astrology and *Zen-Magic* together equals *Star-Zen Magic* and can help us figure out how to decondition ourselves, and how to express the positive energy of the signs in which we have planets at birth.

We have a sun sign, a moon sign, a Venus sign, a Mars sign, a Pluto sign, and so on... and yet... I say we are more than our stars. We would do well to know *intimately* where the planets are distributed in our birth mandala, so that we can work them energetically and weave them into a psychic shield of light energy inside of our own selves.

We are to learn, in a future age still to come, how to work *with* our birth charts. We are not victims of our birth alignments as we did in fact choose our entry chart before we were born here. Yet, what are the potential distortions or built-in challenges we might experience in our birth charts?

Distortions of our true nature through the afflicted emotion known as anger are indicated in the horoscope by the cardinal signs. The cardinal or *beginning* signs are Aries, Cancer, Libra, and Capricorn, and they form the four main quadrants of the wheel of life.

The self-assertion of **Aries** energy, the first cardinal sign, is necessary for survival and holding our own in the world, but when it turns into blind anger, directed at others, Aries energy can cause much suffering. Zen, the spirit of the Samurai, stands with Aries' wish to fight for what's right, but asks the warrior to curb selfishness and ego. Uranus enters Aries in 2011, beginning there, a veritable revolution of the individual identity.

Cancer does not come to mind so readily as a cause of anger or hatred. But think about it. Cancer is the sign of the home and patriotism. How many wars have been fought for the sake of the homeland? How much *angst* – German for fear – have humans suffered over their lives, especially from problems that originate in childhood? Zen, however, sees the Earth as a nurturing mother, giving sustenance to all without concern for race, nation or political belief. All these issues are found in the Cancer sector.

Libra is seen as the only inanimate sign and sector of, "Peace in the heavens", and yet Libra energy can also become anger in the form of open combat against known enemies. Libra is also the sign of relationships, and where, other than in our marriages and close personal partnerships, can we experience more anger? In Zen terms, Libra stands for the desire (we see desire below), for harmony that exists within all – harmony with our environment, our fellow man and with the Tao, the unnameable something that created this universe, sustains it, and expresses itself through its myriad creations.

Capricorn stands for ambition, which when under control, leads an individual to his rightful place in society. But when attached to a self-serving point of view, Capricorn's drive can become a ruthless pursuit of rank and prestige. The Emperor Wu was shocked when he asked Bodhidharma what merit he had earned by building many monasteries and universities? The Zen patriarch answered the Emperor, "None whatever." Pluto is in Capricorn from 2008 to 2023.

The cure for all psychological ills stemming from anger, according to Zen, is compassion. Most of us are capable of limited forms of compassion. We feel compassion for those we love or can identify with. But Zen calls for universal compassion for all beings. The decades ahead are a test of cosmic and global compassion. See how you pass that test.

The fixed signs indicate where an individual may become tangled up in the web of desire, which is greed in its many facets and forms.

Taurus relates to what is *mine*, and to values. It desires love and possessions, and when lived and expressed positively, this then results in well-formed value judgments, that lead to good decision making and steady goal formation. Zen turns our conditional way of setting values upside down. When Master Tung-shan was asked, *"What is Buddha-nature, or the highest value?"* he immediately answered, *"Three pounds of flax,"* – something seemingly valueless. The master immediately saw the conditioning of the monk's mind that led him to believe that Buddha nature was just one more thing to pursue and possess in order to enhance his existence. It is also known as spiritual materialism and has to be avoided at all costs if enlightenment is to be reached at all.

Leo wills its power into being, and seeks to shine and be the best it can be. There is, without a doubt, some ego-centricity to Leo, as well as magnanimity and generosity. But when not *fully aware* of their inner processes, Leos can be too greedy for love, appreciation and acknowledgment. Yet the Sun, ruler of Leo, is also the key to personal salvation – the path Krishnamurti talked about, the one that one must walk completely and with heart. In Zen, when you give yourself completely to life, letting nature act through you, you are said to be a golden-mane lion. **Saturn is the Zen of the Rings**.

Scorpio is the ultimate sign of desire, since its keyword is: *I desire*. No other sign of the zodiac is more associated with the flaws and pitfalls that desire can develop than Scorpio, known really as: *Eagle holding a serpent in its talons...* and Scorpio is indeed known for its intensity and the desire to be consumed totally by the object of one's desire, passion or wanting. The higher level quality of the Eagle then, is the focus that this can bring, for as long as focus can be used in helpful ways. In Zen, this deep interpenetrating focus is known as Samadhi, total absorption of the mind in the practice of meditation, the individual merged with – *All that is and that is not, or the Tao* – the One.

Aquarius keyword: *I know*, represents individuality, rebellion and innovation, and those who seek out the different, the shocking and the unique. Aquarius hosts both Chiron and Neptune until 2010. As of 2024 Pluto begins its trip through the water bearer sign. Positive Aquarian expression is humanitarian and intuitive. A conditioned expression results in people who are greedy for attention by appearing too unique, too special. The Zen spirit is very Aquarian in nature, and enlightenment comes in a typically Uranian flash. Zen masters throughout history constantly used techniques that shocked and jolted their students.

The remedy for afflicted emotions and psychological illnesses stemming from desire or greed is love. Through unconditional love we no longer desire to possess and accumulate things for ourselves, but rather our focus is on what we can give to others.

Ignorance or delusion, and its opposite, wisdom, is the key province of the four mutable signs.

Gemini clearly falls under the Zen description of *the monkey mind*. The sign ruled by Mercury perceives easily both subject and object, but Zen teaches that it is this dualistic way of perception that is the root of all trouble. Rational thinking has its uses, but can only take us so far. It stops short at the gate to true understanding. This is where Gemini has a long way to go to see what its opposite sign, Sagittarius, discerns. Gemini energy loves agility and choices.

Virgo – the other mercurial sign – on the other hand seeks perfection, and wants to know the world in greater detail. But when too narrowly focused, not seeing the forest for the trees, this Virgo energy becomes the *discriminating mind* that Zen seeks to overturn in favor of a more transcendental view. The third patriarch of Zen, Seng-ts'an, once said, ***"The Great Way is very simple. Just avoid picking and choosing."***

The farsighted vision of **Sagittarius** – *I see* – seeks the much broader horizons of knowledge as it studies life and wants to understand absolute if not perennial truth. Its ultimate goal is wisdom, but too often pomposity and prejudice result when one clings too tightly to one's view of the world. In forming our own opinions, we must at all times keep an open mind. Pluto is in late Sagittarius until 2008, transforming these matters. This book is a typical Sagittarius-Pluto product as are the *Special Ones* we will meet later on.

Pisces energy is essentially of pure and unconditional love. It hosts Uranus until end of 2010 while its ruler Neptune is received by Aquarius until then. This reception will produce magnified results in how Pisces will later on perceive reality and unreality. However, no other sign has a greater potential for succumbing to delusion than Pisces. Likewise, the 12th House of a horoscope, associated with Pisces and endings or secret affairs, is the place where the mind either seeks enlightenment or becomes lost in illusions. It is also seen as the place of exile. Pisces is meditation, pure consciousness, and the place where we experience the truth of the words of Hui-Neng, the Sixth Patriarch (638-713 AD): ***"From the first, not a thing is."***

Neptune, ruler of Pisces is in Pisces from 2012-2025, introducing then the most transcendental time ever, and it will clearly dissolve what we thought was the known. The way to overcome ignorance is the active as well as meditative cultivation of wisdom. Jupiter ruled co-signs Pisces and Sagittarius are said to be able to muster this wisdom. Following the Zen saying that, *"those who know speak not, while those who speak know not,"* we must listen for the voice from within and know when to listen to the little Zen monk when he says to others: *"Of course you are right, I did not mean to offend you."*

The planetary therapy of *Star Zen* is to find a way to protect ourselves from these emotionally afflicting energies anger, desire, and ignorance.

I think that we suffer in life because we depend too much on the outer conditions of our lives. According to Zen this is something we can change by meditating on Zen koans, these mysterious verses or riddles to which there are no right or wrong answers. When things are good, we are happy. That makes sense. When life falls apart, we panic like crazy, which is what inner Zen helps us to avoid.

The goal of *Zen-Magic* is to clear away the conditioning of the mind via the study of your birth chart mandala both in transit of time or motions of planets and in our own selves, to find the quality of time and to reach one's true nature.

The ultimate goal of understanding our birth chart then is to reach enlightenment from within. That is ideally what Zen astrology, or *Star Zen* is here to accomplish. One of the Zen koans that best describes the process in a horoscopic energy is known as **Keichu's Wheel:** Getsuan said to his students: "Keichu, the first wheel-maker of China, made two wheels of fifty spokes each. Now, suppose you removed the nave uniting the spokes. What would become of the wheel? And had Keichu done this could he be called the master wheel-maker?" Mumon's Comment: If anyone can answer this question instantly, his eyes will be like a comet and his mind like a flash of lightning.

When the hubless wheel turns,
Master or no master can stop it.
It turns above heaven and below earth,
South, north, east and west.

When Neptune Is In Pisces

2011 – 2025

People are like stained-glass windows. They sparkle and shine when the sun is out, but when the darkness sets in, their true beauty is revealed only if there is a light from within.

~ Elizabeth Barrett Browning

As the world turns in 2011 – this year begins with the moon moving from Scorpio to Sagittarius, with a dose of great optimism as Jupiter (wise ruler of far seeing Sagittarius) moves through the mutable and last degrees of Pisces, conjunct with Uranus. Synchronicity is at work as 2011 opens to a new Cosmic reality; people will have to wake up or go to sleep forever as the matrix veils fall away.

Neptune-Uranus in mutual reception will mean that potentially people will see the invisible world and understand the message from afar. This is a stunning opening gambit, because the ruler of Pisces and Cosmic spirituality – Neptune in Aquarius – is in a 30-degree angle to Uranus, and thus in mutual reception of and to Uranus in Pisces. Uranus is the awakener and liberating planet that rules Aquarius and the masses, which Neptune will have gently dissolved by then, with its own deceptive brand of Neptunian *mind control...*

This alignment augurs well for 2011, and beyond. However, in March, Uranus follows Jupiter again into Aries, and at this point things get intense as we saw in the opening gambit of the decade in 2010. The sign of Aries, tenacious and where things begin, hosts both of the energizing planets that impel and propel humanity toward change, trail blazing and knocking down all barriers; things will have to be new then or die. These two planets are then opposed by Saturn, ruler of order, in Libra, sign of justice and peace; while all are challenged by transpersonal Pluto in Capricorn, urging the earth to restructure, and to ultimately unite as a race known as humankind. This indicates a fight of the Cosmic Titans, Chronos, Ouranos, and Zeus. Based on who is farthest away and thus impacting time longer, Uranus, the breaker of resistance will win...

Asteroid Chiron moves into Pisces in February and for a long time thereafter, until 2017/18; adding to this spirit of merging humans and diluting the artificial boundaries and divisions existing between all the races of humanity. Chiron still travels close to Neptune, enhancing the mood toward a trans-personal altruistic learning curve of mysticism. Jupiter will race through Aries in one year, while Neptune for the first time will have entered its own realm – Pisces – after over two centuries. Neptune will dip back into late Aquarius once more from August 2011 to January 2012.

Memorable May Day may be Code Red: An interesting highlight is May 1, shortly before the new moon, when the Moon, Mercury, Venus, Mars, Jupiter and Uranus (six planets) all travel in the Aries bandwidth across from Saturn in mid Libra. Moon and Mars align with Jupiter on that day at 22 degrees Aries. It will be grand! You will have to get up early to catch that Aries *me-first* display of planetary power. Sun is at 10 degrees Taurus that day, 20 degrees apart. I am not certain the alignment is visible, maybe barely, just before sunrise. The Node on the Galactic Center trines the alignment, so I read this as spectacular and empowering energy, loaded with Cosmic wisdom.

Jupiter in Taurus after early June will bring a grounding energy to all things transpersonal. This means that in June and July, Jupiter and Neptune form a beneficial alignment of 60 degrees, which may well translate into some useful and peaceful, or practical as well as spiritual activity world-wide. I feel the *food chain* will be glad to see all this happen.

Saturn spends all of 2011 in Libra, making certain alliances useful and beneficial. Saturn in Libra is concerned with justice and fairness, and yet the drawback in that position is that it can be too exacting and high-minded, often negating justice. This could lead to a rupture of agreements, or produce unrealistic expectations. Added to that, for most of the time in 2011, Saturn makes demanding aspects to its fellow Cosmic wanderers, the other outer planets in Aries and Capricorn.

Pitted against Pluto in Capricorn, sign ruled by Saturn, when in Libra, Saturn stands no chance; but I must admit the combination spells war, not peace. However, 2011 does end on a spectacular triangular note between Saturn and Neptune, in which spiritual and imaginary concepts might synchronize for a moment and actually yield results. I know this will be a good time to sign treaties or make a new global agreement a reality.

The total lunar eclipse on June 15 is of geometric perfection as it puts the sun close to Mercury at 24 Gemini and the eclipsed full moon exactly on the North Node at 24 Sagittarius, or close to the Galactic Center. There will be a wise moment of Cosmic peace during that day, and the days thereafter. The total counter eclipse of the moon on December 10 puts sun and moon in the opposite degree signs, and is equally well aspected by other planets.

We will see in 2011 that an over-soul type first cause, or original source, is linked by an energy or motion, call it time and spirit, to all the little souls on earth where past, present and future meet in one great gateway to the Cosmos. There will be an upgrading of awareness going on, and whatever happens in the real world, it will be meaningful and suggest a new understanding of ancient and fragmented wisdom. Major discoveries will be made while Uranus and Jupiter are in Aries, and these breakthroughs will change the future of earthlings for the better.

Science will make vast steps toward self-empowerment, and enormous progress to join spirituality in explaining what no religion has ever been able to comprehend or impart to people in terms of what will be achieved. By then, the secret of life may be found. Artificial life and the explanation of what a soul is about is well within sight, but elusively out of reach... 2011 may well be seen in hindsight as the year of empowerment, although I cannot exactly predict what kind of crisis will usher in this new consciousness. Technologically speaking, 2011 is not for the fainthearted – and these little space aliens will have a say in it too.

As the world turns in 2012 – The year of the pentagram, as 2012 is equal five. Many things were predicted for the end of 2012, and I can guarantee you it is not the end of the world, although I am not certain how many of us will be around in 2013. This depends on our stated intent, which we must communicate to the Cosmos. Prayers are useless at that time, whereas setting intent and letting it be known will work quantum physical miracles...

The two main points of interest in 2012 are: First, the second Venus Transit of this millennium in Gemini, after the Transit of June 8, 2004, and secondly, a lunar eclipse between June 4 and 6, 2012, in mid Gemini and mid Sagittarius. We are told that this Transit of Venus inside the disk of the sun ends with the calculation of the Mayans. I would suggest that therefore it also starts a new calculation. We know that Venus, Earth, and Sun form beautiful pentagrams in eight-year cycles. However, to have a pair double pass through the sun's disk is a rare, or once or twice in a millennium type event.

The other alignment of interest, and I use this one to predict the trend of 2012 and beyond, is the winter solstice of 12/21/2012 (equals eleven!) and better even: the full moon a week later, on December 28, 2012. This shows what it really is all about. The full moon in Cancer will have been preceded by a total eclipse of the sun on November 13, 2012 (also eleven, for 11/13/2012 = 11) at 22 degrees Scorpio – tropical placidus calculation. One can state it is by then time to let go of everything that did not work.

The so-called end of the Mayan calendar has a penetrating photo-finish to it. The positions of the planets are simply of a beautiful and sacred geometry with many mutual receptions between welcomed planets sitting in the signs ruled by other planets they aspect, and a great deal of inner meaning.

The Alignment in Cairo at high noon: Sun 8 Capricorn – Moon 8 Cancer – Node 24 Scorpio – Mercury 25 Sagittarius – Venus 15 Sagittarius – Mars 2 Aquarius – Jupiter 9 Gemini – Saturn 9 Scorpio – Uranus 5 Aries – Neptune 1 Pisces – Pluto 9 Capricorn – Chiron 6 Pisces – Black Moon 12 Gemini.

This means a mysterious *Finger of God*, or *Yod* alignment is formed. Saturn and Pluto make a 60-degree angle to each other, in the signs ruled by each other, thus in mutual reception, with the Sun in line of Pluto and the Moon across from it. The base of the 9 Scorpio – 9 Capricorn then points in two 150-degree angles to Jupiter in line of the bull's eye fixed star Aldebaran, which forms the tip of the Yod, or top of the triangle. Mars in Aquarius, visionary, is in sextile to Uranus in Aries, hotheaded, also in mutual reception. Uranus is anchoring a square to the full moon in Cancer and to Sun and Pluto in Capricorn. Saturn in Scorpio, controlling and all powerful, is involved in a beneficial grand water sign trine to both the moon in Cancer and the conjunction of Chiron and Neptune, which finds itself in Pisces; the sign it rules. Jupiter draws no less than seven aspects from all planets.

I do not see why the world would end at this time. I do see here a great moment of reflection. Actually, I sense a certain reprieve and peace of mind, and of course one sees a huge amount of chatter due to the information junkie Jupiter, retrograding in its fall sign Gemini, heated up by

Aldebaran and in opposition to Venus seeking far sight in Sagittarius, and in square to Neptune. Mercury is the only planet that makes no particular aspect to anything, and yet it is positioned in the best spot – the Galactic Center – at the end of Sagittarius, in the 13th sign, Ophiuchus, probably able to communicate wisdom from the zero point field, to the earthlings.

This is the picture we are moving toward as 2012 goes by, and it is the picture we are in as 2013 opens. Saturn in Scorpio is the generic signature of 2012, as is Jupiter in Gemini, which means control freaks will be facing down their fears and lack of control. We want to understand the information we digest, and we must let go of what is unhealthy.

I see 2012 as a powerful and empowering year. Granted, there could be some rather disturbing weather patterns and super storms, as well as extreme unrest when looking at the position of Uranus, Pluto, and Saturn; yet on the whole, I detect a strategic breakthrough for the entire planet Earth by the end of 2012. Dry Land will rise in the *Water World*...

Ruler of the 2012 charts and point person for the many years ahead is without any doubt, planet Neptune, at home in Pisces, ruler of the trident, Poseidon, ruler of Atlantis; also holding sway over oils, leaks, super storms, the raging seas, illusions, dreams and deceptions, things mystical and invisible, yet empowering...

This planet will play its tricks on humanity by 2012-2013, as it is beginning its long journey to rediscover the lost civilizations of the continent that sank some 13,000 years or a half Great Year cycle ago... while assisted by Pluto and Saturn in mutually beneficial signs and positions. This enables solutions for finding the hidden treasures – from the unified field, the zero point field, to solar wind energies from deep space, all the way to scalar and orgone technologies for liberating and empowering the masses.

2012 and 2013 will see extreme challenges and the type of energy shortages that will have devastating effects on every aspect of our world. All infrastructures will be harshly and persistently impacted until 2020 and even beyond. A world intelligent-management organization to regulate production and distribution of existing resources, based on a search for alternative, renewable sources will become an absolute necessity.

Behind the scenes, Saturn in Scorpio, Pluto in Capricorn, Neptune in Pisces, the managers and magicians of a future world will assist this chain of events in a manner that will restore equality to the world's peoples and indigenous races. It will have to be situated beyond the grasp and control of special interests and dominant powers, and it will have to ensure fairness. This is a tall order, and yet it is what 2012-2013 will be about. This new type of justice will have to seek as its instrument of world stewardship a fair force for global confederations. This is what the transpersonal planetary positions show us, into 2020 and beyond. Then one could say that the memory of all would – metaphorically speaking – be erased.

With Neptune in Pisces and in constructive aspects to Pluto in Capricorn during more than a decade, the spiritual structural element of human evolution is strongly integrated to the whole theme of global energy resources. An extreme acceleration of all of this will occur, once a majority of humans realize how empowered they are. The ruling elites will have to be replaced by the Indigo children, such as the future new leader of India, who, in 2004 at age 33, is on the horizon.

The human migration of which I wrote in my chapter on the *Transmigration of Beings*, is a Cosmic plan in motion. Humanity is evolving on one level and migrating on another. In the sector of its evolution, humans are becoming more technologically advanced, with the ability to multi-process more sophisticated visual, aural, and intellectual data. The brain system is changing to become more holistic in the way it will process information. Computers play a major role in this evolutionary track; yet, until and unless humans learn again to see with and from their heart, as they did millennia ago, the evolution is just that – technological, or Aquarian.

Humans are also migrating from the notions of divisiveness – separation by means of nationality, religion, etc. – to a grand unification through spiritual Cosmology. Humanity is eventually coalescing. Even Nostradamus predicted the races would mix to *one*, although it may not seem like it because we will continue to have wars and conflicts until 2016/17. The evolution is happening gradually. Over time, by 2020, the world will be a better place, after the dust will have settled...

If all of this means a pole shift or tectonic super quake is due in order to unearth the knowledge and knowing that was once the unified science of our star trek ancestors from Mars, Aldebaran, and Sirius, then so be it. I invite you to join me in the discovery of our origins... back to the future.

A united or a blown-up world is approximately another Venus Transit away – in late 2020 or thereabouts... when wise man Jupiter, master of time Saturn and lord of nuclear power Pluto conjoin for the appointed grand board room meeting at 22 degrees Capricorn. That is Saturn's or Sauron's fiefdom, sign of the horned one... The good thing then is that Neptune is located only 60 degrees away in Pisces, gently maneuvering the situation toward an esoteric or metaphysical understanding and peaceful outcome.

2021 – another 5 year – and nine years into the future as seen from 2012, begins with both Jupiter and Saturn in Aquarius, adding to the note of transpersonal quests toward unified humanity. Once again a new President will be taking office on January 20, 2021 – if the US is still functional as such by then. The election, if it took place in November of 2020, would have happened when both Saturn and Jupiter were in Capricorn, Tecumseh curse; the same as was the case when John F. Kennedy was elected 60 years before or in 1960.

Make of this Chinese cycle of time, 60 years, five elements, twelve signs, five Jupiter and two Saturn returns what you wish, but Nostradamus said there would be a time when the new land would meet its demise. I think 2020-2021 is when the real new cycles begin, with Uranus in Taurus and in square to the Aquarian bunch of planets, disposing of the lot rather rudely. It could simply be that people misplaced the digits when reading the Mayan Oracles. 2012 or 2021– which will it be?

As the world turns in 2013 – The first breath of the cycle. 2013 will be known for its watery climate and a grand annular eclipse of the sun on November 3, in Scorpio, when the sun will be osculated by the moon and by a retro Mercury on the North Node and in line of Saturn – and all this precisely in the region of the Southern Cross constellation at 12-13 degrees Scorpio. This is known in Sabian symbolism as, *"The Cleverness Of An Inventor Experimenting,"* suggestive of an innovation type year. It will be a time to let go of the old and what doesn't work. One could say 2013 will be the year of the Southern Cross.

Jupiter will be racing through Gemini, ever the information junkie, with lots of superficial chatter, short trips and fun travels, until the end of June when Jupiter will go crabby in Cancer for the second half of 2013. A pretty alignment of a grand trine triangle in the three water signs is made all through July and August between Jupiter in Cancer, Saturn in Scorpio, and Neptune in Pisces, so the spiritual dreams can cautiously become a reality and merge. Jupiter in Cancer then has a tendency to become mushy and overly security orientated as we witnessed in 2001 and 2002, around and after 9/11.

Hotheaded Uranus in Aries plows ahead with an extreme vengeance, blazing the trail, inaugurating new ideas, breaking down all barriers; all this in extremely challenging positions as it anchors the double square to both Pluto in Capricorn and Jupiter in Cancer. In other words, Jupiter will oppose Pluto in August, and the sharpest alignments are at the end of July when Jupiter aligns with Mars in Cancer, and around 8/20 when the T-Square in the cardinal signs is perfect.

The Russian born, *Boriska boy* from Mars mentioned 2013 as the second year – after 2009 – when a large catastrophe would occur to the earth. End of August will certainly be a challenging time, and a tectonic type of shake-up is possible or probable. Black moon is then also in Cancer, opposite Pluto. Summer 2013 is a time in which people will look for security within their shells...

Despite beneficial and mutually receptive alignments in the water signs, the other aspects will create a strategy of tension. However well sustained by the best of intentions, it will lead to confrontations and silent conflict. Saturn in Scorpio acts in a controlling and subtle manner, and secrets will be amassed and hoarded.

Spring 2013 will be memorable; March 27 is a splendid full moon in Libra pitting that moon against Sun, Venus, Mars, and Uranus in Aries, to make this an extremely powerful and surprisingly super-charged start into a new *Celtic Year*. Take in the first breath of fresh air of the new cycle by Mayan accounts...

2013 may well be the year of a profound upheaval such as earth movements and social incidents of global magnitude. It will make the 2004 tsunami look tame. The type of thing that would make it appear that the whole earth is shaking. I would expect 2013 to become an emotional type of year, or subdued; one fraught with issues of safety and overly exaggerated security concerns.

In my opinion, the many water planets will make the latter part of 2013 potentially a stormy and wet year. Watch the floods and watch out for water world... with Pluto in Capricorn, dry land is high up or in outer space... I feel there will be a backdrop of revolutionary electrical undercurrents in 2013; however, these will only come to a peak one year later, when by the total eclipse of October 8, 2014, Uranus in Aries – squared still by Capricornian super-controller Pluto – gets a sustained fiery grand trine from a lion-hearted and managerial Jupiter in Leo; and a rash if not reckless Mars in Sagittarius. Mark thus the end of 2014 as the moment or passage when action will be happening and new developments will unfold with bravery and true courage.

2014 – A New Cosmology Discovered. 2014 is at the mid-point of seven years from 2007 to 2021, the fourteen-year span that I characterize as the core of our much needed *passage to peace*. 2014 ends with Saturn moving into Sagittarius and ruler of Sagittarius, the central light of the galaxy, Jupiter in Leo. We move within one year from a water to a fire grand trine, thus in a way constructing a tetrahedron through the skies, marked by powerful outer planets.

12-12-2007 is when Jupiter meets Pluto in the Galactic Center at the end of Sagittarius, which we can understand as the core of wisdom; a seemingly black hole out from which emanates everything that creates life. By then, I wish to see the beginning of a new cosmology emerging.

The year 2014 promises to be the all-out informative and gruesome as well as breakthrough and compassion-driven year. Everything imaginable under the sun will be experienced in that core year. The fire, water and earth planets form aspects to one another that can only be interpreted as freedom from the known. The grand water trine in early 2014 between Neptune in Pisces, Jupiter in Cancer and Saturn in Scorpio means that compassion and perception are at a rare high for all of humanity. We will all pass through the grand portal of existence at some level of our being. The later fire baptism grand trine made

399

between Uranus in Aries, Jupiter in Leo and Saturn in Sagittarius into 2015 impels humanity finally to break through the boundaries.

Beyond 2014, if I care to add another Uranian seven-year cell renewing cycle, I wind up in 2021, with most planets conjunct in visionary Aquarius, challenged by Uranus – by then in Taurus – promoting the age of the *cosmic electron*, or a new cosmology. An entirely different understanding and knowing about where we came from will set in during the years from 2014 to 2021.

The history of the universe or origin of the Cosmos as we understand it during the crossing taking place between 2014 and 2021 is what the term Cosmology is all about. How can we as human beings comprehend and esteem what we are, if we do not know where Cosmos emanated from? On the other hand, how can we understand Cosmos if we cannot even evaluate our own little selves? The question is who created the creator? 2014 will change all of this comprehension. It will be resolved in a satisfactory manner by 2020 when all planets of great importance: Pluto, Saturn, Jupiter, Mars, etc. are lined up in structural Capricorn.

My own cosmology is simple, or at least user-friendly. I submit that out of nothing, nothing comes, to put it bluntly. Often it is more important to know what we do not know, rather than that we pretend to know anything about anything. At least we can be certain about what is *not*, rather than what is. How can anything have exploded if no-thing existed? We do know therefore, by simple logician's deduction, against all odds, that the *Big Bang* theory is rubbish, absolute nonsense. Cosmos is a self-conceived and self-organizing matter, or concept full of currents and motions. One could term it the electronic universe or the – *Origin of all that is.*

The most graceful as well as merciful thing on planet earth is the apparent inability of the mind to coordinate and correlate its contents in a useful or functional way. It is this flaw that makes my invisible extra-terrestrial friends smile at most human-beans. The belief based sciences, each straining in its own dimension, have hurt the earth; but someday the piecing together of the diluted puzzle of Cosmic knowledge will open up greater and more functional transformative realizations of reality. As humans we will either go insane from the revelation of knowing, or we will run away from the shadow light into the serene peace and quiet safety of a new dark age.

An interdisciplinary quest for truth or the humanistic inquiry into cosmology such as this, demands that communication itself must become absolute, or else it can pose a challenge. The most important difficulties in communication manifest when one is questioning something already *known* to be true. On such issues of underlying principles such as, "Where did we come from?" the confidence behind established – belief based – ideas can be so high, the discourse becomes almost senseless. This problem is furthermore highlighted if not aggravated by the fragmentation of the process by which information is gathered and evaluated.

The last millennia since Atlantis, and the past few centuries after the *age of enlightenment*, have brought to humanity a dissection of knowing. So-called sciences have become incompatible with each other, and the detailed specialization of almost every spiritual or intellectual inquiry emanates risks when assumptions within one

401

discipline are based upon prior assumptions in other fields of knowledge. And yet, without merging it all, feeling, knowing everything to its core, we are lost. The human brain seems – but is not – incapable of connecting the dots of all that matters. What matters is the executive summary, the mother-load of all knowledge's and disciplines.

We *think*, but we do not know for a fact, that nobody is knowledgeable about everything. And when considering possibilities outside an expert's personal expertise – whatever that means or does not mean – it seems only human or at least natural to defer to what specialists in other subjects claim to understand. Would you question a brain surgeon? Would he question a rocket scientist?

Who is to say we cannot comprehend all that matters, and what are the consequences of this questioning and comprehending when theoretical suppositions, such as the Big Bang, albeit perceived as fact, cannot account for compelling new data, such as Planet X? Do you think these astronomers have any clue of what they are talking about? What we must understand is that everything always changes, and this transformation of perceived knowledge is constant and ubiquitous. In the end we know very little – and it is all right to admit it to ourselves.

Given the extreme fragmentation of this traditional house of Babel type of established science community today, it seems impossible or virtually unmanageable to draw a new image of where we came from. It seems equally impossible to imagine that the enterprise of Cosmology could ever bring together and correlate its contents into one grand multi-dimensional canvas of constant knowing, that makes sense to everyone.

I was called the great simplifier, but in reality I am an extremely complex person. The only thing that makes me slightly different from most is that I know I don't know a lot, and therefore I listen and comprehend. We will all undertake extraordinary strides toward that day after tomorrow, which was envisioned by people such as Ray Bradbury, Immanuel Velikovsky and H.P. Lovecraft, when the things I predict or foresee will prove feasible via totally evolutionary approaches in how we comprehend our world.

The Cosmic ether, or magnetic-electronic phenomena will receive the full attention they deserve, once we start looking beyond what it is we believe we know. There is an underlying coherence to what we seek in Cosmos. To those who identify unifying principles, the new horizons will be both breathtaking and inspiring and also of childlike simplicity to comprehend.

We need no institutions or mystics to teach us where we came from, or where we are destined to go. We are mutant memories connected to the Cosmic mind: all knowing, all powerful, and self-organizing. The focus of choice is inherent to the human DNA. No need to pray; it remains in your power to decide which way you will go.

This book – as part of a *Zeit Geist* or *Time Spirit* – represents a view of what we want to know about, pulling into profoundly deep and at the same time crystal clear focus what we feel we cannot see. This is a view of the physical and psychic universe, a vision from sub-atomic particles crossing into galactic realms. For those visionary few who are clairvoyant and sufficiently able to do their own psychic remote-viewing life is the pathless unknown.

The horror which the elite in their fearful state of mind, and their think-tank underlings in their experienced certitude feel when they realize their power of knowledge is a total illusion – is to you a form of freedom. The first rush of uncertainty, when ideas long taken for granted are thrown to the wolves, or at least thrown into question, by sure facts and simple reasoning will divide humanity. Some will continue to believe in miracles, and other will understand.

By 2020, the science establishment and political house of cards will perish, as did the Inquisition after Galileo posed a few simple queries. With the courage to see clearly, the adventure itself could well be the most graceful thing in the entire universe, by which we will be adding new insights to the greatest dramas of early human history. We will also be providing a vital perspective to humanity's actual situation in the universe.

As we enter this portal of knowing, all things open up grandly. In fact, it is not fear that accompanies us through the doorway to the Cosmos; instead, it will be the joy and the adventure of discovery. It is the anticipation and expectation of finding what is *beyond the beyond* that will propel us to better comprehend who we are.

Where did we come from and where are we headed? Beyond the point of 2020-2021, with outer planets in grand alignments in the transpersonal signs of Capricorn and Aquarius, the switch will happen from structural to electronic comprehension and a dynamic interaction that is not tied to the known. Free yourself from the perceived reality, because the truth is never what you knew or know, it is always and forever new...

Dawn of The Time Lord

I wanted only to look back and say: "There! There's an existence which couldn't hold me. See! I vanish! No restraint or net of human devising can trap me ever again. I renounce my religion! This glorious instant is mine! I'm free!"

~ Paul Atreides, Dune

You might imagine that a time lord who was able to appear in this space of our fragmented existence, would impress upon his own self from the past, that he exists... And you might imagine that he would use his presence to influence his own progression across the invisible bridge of time. But you are wrong.

As a child I naturally loved to go to Chillon castle, which often drove my parents mad, as they did not want to go there as often as I demanded. From an early age the castle was my home, if not my allowed home. But I have no memory of meeting the *Master of The Light* there. The wise do not impose their will on others; they nurture and wait, leaving the growth to happen naturally and without effort.

I did see and understand many warnings pointing at our collective future, but those dreams were my own ability to perceive and had nothing to do with *influence* from outside. The dreams and warnings I experienced in my childhood and which continue even to this day are now slowly beginning to manifest; my visions are becoming a reality. I have always said I wanted to create solutions based on my visions. I am a great optimist and I know the human spirit will come through, but only if the spirit is fed the correct information of the future – and only if humans learn to adapt to this new future.

A number of friends have also experienced psychic warnings and dreams since childhood. So, I am not alone with my knowing. Magically, we have now met, and have formed a loose network, as the caretakers of future earth have closed the global circle. Together we are developing a plan. Its essence is to adapt to the changes facing us and not interfere in the natural flow of events.

The message I am now receiving is of the utmost importance: The guardians have told me we have to adapt and in doing so save our world. We need to build what I call *Inter-Dimensional Schools of Light*, or centers for contact, throughout the world. We have to move beyond *survival* and into living life to its full capacity, which also means to cooperate with each other and end all conflict now.

The centers of world learning will be spiritual schools that transcend all restrictions of religious and political dogma. Humans must develop spiritual muscle power, based on freedom and free will in order to adapt to the coming earth changes.

To survive physically is not enough. This has absolutely nothing to do with belief systems, for I believe in nothing. It has everything to do with knowledge – ancient Cosmic knowing and the alchemy of true science. We will live together once more in balance with the earth and its inter-dimensional guidance systems.

I see us as spiritual guardians of the earth now and of the future earth. We are all here specifically to make it possible for the future to survive. Nothing less is at stake than the survival and future of humankind, which our so-called leaders have endangered.

At this time more than ever in the earth's history, we need contact centers where people can connect their own inner world with the real inter-dimensional one we live in. Inner contact is the command center of all, where we each make the decision to meet life or run away and hide. Life would have no color or variety if not for the unexpected challenges that appear, there is nowhere to hide.

Cosmic contact means to learn be in command and move in power. This is something that frightens most people, because to move in power means to let go and move with the flow. Learning to command the contact from within ends the fear and the isolation. Nature and spirit guides will also guide us through upheavals, if we know how to listen.

No one knows yet the true nature of the coming changes. Focus has been on physical changes to the environment and eventually to the earth itself, and few people realize that the most major change will be about what happens to and within the human psyche. The landscape of the human psyche is much larger than any planet or continent.

When the psychic capability changes, this alone will come as a shock to many people unless they are prepared to deal with it. In reality we must learn that the psychic world is more real than the physically manifest one. The contacts centers will be part of a self-discovery training that will prepare the mind for changes or shifts in perception. The key foundation for this training will focus on learning to make contact with one's state of being in the future, then using this contact as a guide.

Contact is with the inner core of human existence, the *C-Enter* of being. This is the ultimate contact *C-Enter* for self-fulfillment. At these places of learning people will exchange ideas. Then they may go back into their worlds, apply what they experienced and teach it to others.

We are entering some sort of spiritual *Special Forces* type operation at a time when we must have no fear – not only from psychological operations run by the *New World Order*, but also from the alien run matrix. It is a time when we must discern truth from lies and falsehoods. We must be able to act with inspired faith and remain positive, no matter what transpires outside. Whatever happens in the next few years, we cannot and must not let it effect us negatively. We have to have grace under pressure and act with courage. Inner psychic recall helps to circumvent this human type emotion called fear.

With psychic recall we know exactly what is ahead. We can no longer stand by and watch things fall apart around us. Operation 9/11 was an exercise in deception perception, a showcase of a carefully planned mass-influence psychological operation. There will be many more such games. We need to learn to use our psychic gifts, especially since Uranus will spend seven years until 2010 in Pisces, to reach beyond the natural into what normal people call the super-natural, or what I would simply call the divine. All of us can be prophets and teachers. And when we are in this state, we can also change the course of our lives and the outcome of events.

The system of our current society is a hierarchical system of control that hides the truth behind convincing lies... clever manipulation by the few, of billions of people. This deception has been happening for centuries. However, since 2001 the attempted control has become deadly serious.

Most people are inside the matrix and are happy to live under the conditioned mass hypnosis. The majority are content to have other people tell them what to do, how to live, what to think, and even what to believe in. I urge you: Do not believe in anything! However, I advise you: Acquire knowledge!

Examples of mind control are: religion, education, society, marriage, the media, radio, newspapers, economy, jobs, government, food and *medicine*. Maybe you were not taught to take responsibility for yourself and take charge of your own life. Or possibly you never cared to. You may not have questioned the establishment and simply did as you were told because you could not be bothered, or because you didn't want any problems.

You feel comfortable and safe the way things are. It is terrifying to let go of what is familiar and seems safe. Besides, you enjoy tradition because it gives you something to identify with – your identity. Perhaps you prefer to be blind. Throughout history, many have even teamed up with the rules and rulers of the game and their players, to manipulate and control everyone around them. They have sold out for family, fame, money and power. You try to adapt and not cause any ripples – right?

Perhaps you do know the truth but don't know how to opt out of the matrix. It is all about choices and consequences. Decide! Become free and powerful in a way you cannot imagine, or stay in the system and live with the consequences. Religion is divisive and has been one of the easiest ways to control the masses. Throughout history, men have used the deity to trigger a belief system that keeps them enslaved. These mind controllers take their sermons from ancient books that have been stripped of references to free will and cosmic or universal truths.

The holy books consist of stories that have been passed down through the generations, tailored with each new edition to suit the social and political needs of the time. Male scribes and a few other elitists were the only ones who could read and write at the time these stories were collected into scrolls or books.

Divine inspiration exists without needing to be organized. Each of us is part the divine cosmos, something that has nothing to do with organizations. You are an evolved soul that came here in this lifetime, on this planet with one message: Free will – use it as you see fit. But remember, how we behave toward others reflects back on ourselves.

Love is the underlying power of Cosmos. When a powerful person with charisma speaks, people tend to listen and blindly follow. Listen to yourself! Be your own leader. Listen to your heart. Follow your feelings and walk your own path. Harm no one. Do you need to be dependent on total strangers to guide your life? Do you need to worship statues and symbols? Do you need a specific humanly-made structure such as a church, synagogue or mosque in which to perform rituals based on these dependencies?

In Frank Herbert's Dune, Paul Atreides becomes the legendary Muad'Dib, the leader of the Fremen and controller of the spice on the desert planet Arrakis. Foreseeing the destruction of mankind, Paul adapts to moving the future onto a new course and away from the destruction of the species. He succeeds in altering the course of the future by refusing to be the messiah, thus freeing future generations from the most oppressive tyranny the universe could know, one that would be created in his name after his death.

Paul's son, Leto Atreides, begins to understand his father's vision. He too adapts to turning events away from the inevitable death of the species by orchestrating the process that brings about his own destruction, thus freeing mankind.

A key consideration that runs throughout the Dune series is that the greater mass of people are as responsible for the ensuing mess and danger as the controllers and manipulators – because they do not want to take control of their own lives.

This is where the real problem begins and ends. Do you feel you have to believe in something? Do you not see that all of this leads to division, conflict and unhappiness? How have leaders and organizations become so powerful? Who gives them that power? Why, after thousands of years has mankind remained violent, divided and in conflict?

While fighting the Romans, in the Highlands of Alba (Scotland), the Chieftan of the Picts, Calgacus said: *"To plunder, to slaughter, to steal, these things they misname empire; and where they make a desert, they call it peace."* The Picts were the last of the true free people of the Northern hemisphere, and with the loss of their land they later faded away into obscurity like the elves.

Today, the people who have inherited the earth are moving ever closer to complete self-annihilation due to complacency, an inherent laziness and lack of ability to act for themselves. The monster of authority grows ever larger and heavier due to the weakness of those who do not want to take full responsibility for their lives.

The Universe does not judge or condemn. Creation is energy, which cannot die, it can only transform. By the natural cosmic and planetary laws of transformation, government and religion are on the way out. If you continue to depend on an outside agency to co-ordinate your life for you and the system collapses by default... What will you then do?

"A man must be half mad to imagine he could rule even a teardrop of that volume." – Paul Atreides, while looking at the stars through his eyeless vision, from Frank Herbert's, Dune.

Cosmic Star Navigators

"Astrology is a science in itself and contains an illuminating body of knowledge. It taught me many things and I am greatly indebted to it. Geophysical evidence reveals the power of the stars and the planets in relation to the terrestrial. In turn, astrology reinforces this power to some extent. This is why astrology is like a life-giving elixir to mankind."

~ Albert Einstein

The war of control over the consciousness of planet Earth is being waged now and until 2020. It is a spiritual war, a struggle for the control of minds, the resources of the earth and in that respect human beings are seen as a *resource*. What is important in the years ahead – as we enter the wave of new energies challenging the hierarchy – is to integrate our actions with the new brain and move away from old brain fear based survival instincts. The ancient trauma is now behind us as we envision an integral, non-fractured future.

There is nothing in the *illusion of self* that needs eradicating, because it is just that: An illusion. It has no substance, other than the substance you give it. Just let it go. Only then can we come more fully into relationship with powerful aspects of our experience of existence, and we can alter our perception to *see* and understand the value of the experience of life, and be grateful for that which completes our experience of the *Grand Mythological*

Vision – the one that is the envisioned integral future of mankind, beyond trauma and beyond the fear based mechanism of self-survival.

All things considered the fact that mankind exists in this Cosmos is actually a miracle. But then I have always been of the opinion that miracles do happen, especially once we intend them to happen, which is to live direct action. In 2020 the outer planets form a precise configuration in Capricorn, its lead planet, Saturn among them. This is indicative of a phase when leadership by example and serious self-sustained community building, for a period of timelessness, is demanded.

In the years ahead we will begin to create functional networks of cooperation that can meet the coming earth changes. The effort to translate economic success and modern investment banking into a stable process of growth – rather than follow the roller coaster ride of a doomed economic environment – requires a healthy and sustained insight into the forces moving society.

Company ethics and ecological integrity investments are not enough to ensure future success. A vast majority deny the reality of these changes. During my retreats I show my clients the understanding of these forces and how they are changing the world. All this will then form the basis of future knowledge to build a saner world. The major steps are awareness, knowing how to better use and move resources, knowing how to meet a changing economy and how to correctly estimate the outcome.

In this respect, I am showing my clients that knowledge is its own direct action. Those interested in ecological businesses and integrity banking, who are trying to ride the waves of the system must understand that the entire system is about to collapse. Many years of careful investment could be lost overnight. In the light of this fact, a business investment carefully structured to create a better world, will be lost in the general economic and social turmoil.

You may feel safe in the knowledge that your lifetime's work is placed with intelligence into a system that badly needs restructuring. However, the hard fact approaching us in the years leading to 2013, and beyond, is such that the economic social collapse will not discriminate between your resources and the resources of multi-national corporations.

In the future there are no businessmen and businesswomen; there are economic artists and caring architects who co-operate through a network of pooled resources and integrity type venture capital. The future business world is the creative and innovative realm of economic artists and innovators who can understand and work with the new template structuring society.

Economic high-priests and financial gurus will no longer supervise the organization and structuring of societies resources because they will have been sent packing; instead the intelligent utilization of earth and human resources is a matter of highly precise architectural planning and investment design.

The *Cosmic Architect* is a very ancient and powerful concept. The sacred geometry of economics has not yet found its way to the surface of the modern human mind. With a return to this shamanic knowledge base, the science of economic management becomes the playground of the artists, the creators, and the initiators. To create and maintain social order, *cosmic design* must be initiated into the geography of the human landscape.

Energetic engineering is an inner configuration that on the earth would be described as health. The ancient systems of Yoga and of the Healing Tao are perfect generators of energetic engineering. You have been led to believe by this society that health comes and goes by chance, that those who have energy are lucky and those who do not have energy are unlucky – as though mysterious karmic forces are building or withdrawing the energetic life force of a human being. The ancients knew differently.

All of us living on this earth are ancient beings. Our souls are timeless and eternal *Cosmic Star Navigators*, the stars are the reflected diamonds of the soul. Unrestricted by time, we are time travellers. We inhabit and experience different realities across time and space. Today's *European* is tomorrow's *Iraqi*. Where will your own soul-eyes awaken and open within the next two thousands years of earthly existence? Who we are in any one incarnation creates who we are to be in the future. That is the definition of health. Your decisions and actions today create your many tomorrows...

If you imagine a highly advanced being, who is you to be many lives from now, looking compassionately back at who you are now... what guidance would you give yourself from the future? More importantly, if you would listen to that guidance, how would your life change now? The ancient systems used in China and India, were part of a sacred *Cosmic Science* designed to connect the user to the wisdom of the stars and beyond: The alchemical soul-science of the *Star Navigators*.

You are the energy of the stars brought to earth.

Energetic engineering is the conscious involvement and fine tuning of who you are across time and space. This is how the star beings define health.

In the very near future we have to start thinking about how we will live in the most essential, basic and down to earth physical ways. There is a need to consider topics of intelligently structuring the logistics of a home, the resources of your shelter, the growing and storing of food, the back-up of power, energy and clean water.

Retreats and centers for *Cosmic Survival* will connect with experts and specialists on issues pertaining to strategy and protection. Transiting through the Cosmic changes of the next decade will potentially entail long temporary power outages and the erosion of coastal communities, the collapse of inner cities and of social infrastructure. We will also see the loss of entire regions of once inhabited areas.

We need to find new and peaceful technologies to be used in order to survive independently and to make a functional transition away from an obsolete system. We will learn the communication from the future by leaving our past behind. The *World Wide Web* will experience a shift toward an *Omni Net* where wisdom and brand new computer languages and open source software, as totally new operating systems are introduced.

Dynamics of spiritual survival will include the ability to develop clairvoyance or the art of seeing. In order to see the correct future, you must acquire the capacity and the skills of understanding the signs of warning and use this ability to bring help to your friends and associates through times of turmoil.

There is great need to learn how to be a functional

community leader from whom people can seek counsel. People will have to learn how to communicate with nature and how to read the silent language of trees, rocks and bodies of living water. Life in the years to come will develop along the lines of entirely new concepts of innovation and of creativity. Only the skills of adaptation and unfettered inventiveness will help resourceful communities of the future to take charge and to restructure a meaningful existence for all the people around them.

With the establishment of various retreats, we are able to advise the free people how to better adapt within the new parameters, in a world under siege. The key to success is a sovereign and completely autarchic or self-sufficient structure that is in harmony with the environment, whose laws are Cosmic in nature, i.e., superseding human laws.

Dr Edward Bach, the creator of the Bach Flower Essences, was one such innovator of new methods in a new world. His genius was applied to helping people regain their balance as a basis for healing. Dr. Bach went beyond the usual symptom-drug matrix to search for and discover a more powerful way of strengthening the psyche and dealing with the problem-maker (within us).

The Doctor Edward Bach Foundation continue to apply his vision: "Dr Bach worked for several years in hospitals and was well aware of their negative effect on the human spirit. After leaving London and starting his work with the flower remedies he dreamed of a different kind of hospital, where people would go freely to find themselves and learn the lessons their life is teaching them. He dreamed of doctors who would understand people as individuals and study human nature rather than test tubes and lab results. And he imagined patients taking charge of their own health by understanding and accepting the needs of their souls, rather than attending to the needs of the body alone.

"When you think about these things you come to a startling conclusion: the hospital and the doctor and the patient that Dr Bach describes are all the same thing. They are all in each and every one of us. The hospital is not a building somewhere, but a state of mind inside us, an angle of the soul. The doctor of the future and the empowered patient are you and me, each of us helping ourselves and each other with these remedies."

In the years ahead totally new methods of healing with resonance frequencies, sound frequencies and light will be discovered and applied. We will discover that molecular interactions of essences are profound forms of communication, as well as being a source of physical and psychic renewal. The essential communication is balance. The interaction of physical substances is its own subtle language. The physical carries a magnetic signal encoded as a light frequency. When we interact, we communicate. Once humans understand this, they will no longer destroy this beautiful world. Healing is the art of listening.

Power of Dragon Elves

"We, the most distant dwellers upon the earth, the last of the free, have been shielded by our remoteness and by the obscurity which has shrouded our name... Beyond us lies no nation, nothing but waves and rocks..."

~ Calgacus Pictish chief

The Master of Light came to meet me one evening in winter of 2006 in my own modern place, overlooking the lake from even higher up than the castle of Chillon. In fact, we looked together at our castle of 800 years ago, and we felt this was enough to understand the story of the two towers. Some say 2006 was 1706 because the Romans had falsified three centuries out of the documented time zone passage. "Is that really possible?" I asked him.

He stood in my own sanctuary, quietly looking at me, and he gave me a dragon-elven explanation about the things I had pondered and considered since I was a young boy. "You know all this from *before* – and you can activate the knowledge just as you activate fluid intelligence," he indicated to me with wordless thoughts.

I heard him say, "Everything you observe, look at, take in, see, think, say, write and do is who and what you are. We are all of Cosmic mind. All of that you is a mirror of you – your tomorrow looking back at you now. The language is of starlight, not of thought, not of words known to earthman from the past. The imagery is there upon demand. Call on it any time. You can even go back and alter the time line. With starlight you change the impact of perception."

The Master of Light suggested, "Separate your thoughts from the thoughts of your mirror and from the thoughts that are actually not your own." In that moment I became aware of the reality that thoughts can be and are beamed into the brain from many sources.

But how can I know the difference between thoughts that are my own and those thoughts coming from outside? The answer came to me, as though the question was the answer, and the two were Cosmic twins. Ask the right question and the answer is part of the question. It was, as Krishnamurti always said: *"You have to ask the right question."*

I wondered if Krishnamurti was also a Dragon Elf of the Shining Ones. Which thoughts are my own, and which enter the mind from outside? As though transported above the lake, I was aware of the mountains and the sky. There were no clouds and the lake was shining blue and gold in the sunlight.

Thoughts are part of a genetic memory, the mountains said. The memory and the thoughts have been turned on mankind and used against him by slightly cleverer entities, but that is soon coming to an end. Humans were vulnerable because they live only in thought and not awareness.

As soon as you are aware, you immediately see which are your own thoughts, and which are the manipulated thoughts of the genetic memory being used against mankind. How else could they control us? They had to use a doorway into our worlds, and the most vulnerable doorway was linear thought. Awareness is our protection, and as few human beings on this world are aware – the result is the world of conflict.

I was aware of being at the same time in my place and yet seeing from far above the lake. I could shift between the two, but when *The Master of Light* spoke I was once more back in this reality. He said, "The space in between is where you are now, both in birth time and place, at any moment. Now fast-forward it to where you wish to see and be. What and who do you see? You are a mirror of seeing to both of them."

He pointed me to a small black silk bag ornate with golden dragons sitting on my desk. He asked me to choose a rune from the bag. They were the runes made from the talking trees of the elves, from the Celtic world of the Scots pine tree. A branch had fallen the day I walked into the ancient wood, and that branch became my runes. The Celtic tree had talked to the stars for over a thousand years.

I reached into the elven-rune bag until I could feel one piece fit into my hand. It was the diamond shaped rune, *Ingwaz* that prompts you to enact the now in the present form. "Think about it," he advised me and left my vision, "We shall meet again. I shall appear when you need me."

The star-language of the runes are light-codes within our DNA, but at this present time those light-codes are sleeping giants in the caves of obscurity, where they are hidden and protected until such a time that people will once more awaken to their true soul-origin and in wisdom use this power wisely.

Deep in the earth the dark substance we know as black-carbon is transformed by crystal light into a bright diamond. Until discovered, the diamonds remain embedded in the earthly black of inner Cosmos. In the same way the light runes lie embedded in the sleeping segments of our DNA.

I once more recalled the recent vision of the sun: "The son of God is the *Sun* of God." Much later, in Scorpio moon, with the Sun 00 degree Pisces and with Mars 00 degree Gemini, I received telepathic knowing from the early morning rays of the sun. Translate Christ as crystal and you have the crystal sun, or the crystal light of the sun. The children of Christ are the children of the crystal sun. Not a man, not a human being, but a being much older and much more capable of directly sharing the *light of God* with mankind.

No more need for churches and temples (which block out the light of the *Sun*), no more need for spiritual interpreters... as the light cannot be interpreted, but only received directly. The faceted crystal is the prism through which the rays of sunlight reflect as colors of the rainbow. From which we have the rainbow tribe of light, or the rainbow serpent.

Our light encoded DNA is the rainbow serpent, or dragon gold. Our consciousness is the crystal, the diamond transmuting sunlight through the many faceted prism of the soul into the beautiful rainbow colors of reality – our world. We are the children of the sun, the children of light.

The dragon elves are at times referred to in other languages as the *shining ones* coming from an advanced light race, to guide humanity through extraordinary changes in genetics and into a time never experienced before on earth. They are connected to a Cosmic knowing system beyond what any sage on earth has ever been able to tap into. These races withdraw at times into another dimension when brutality and cruelty becomes too much here, as is the case now.

They withdraw to await their appointed time. If really there is such a thing as a *second coming* then it is the return of the *High Elves*, partly in form of hybrid humans. They will not return until they see the symptoms of true knowing in a critical mass of people on this planet. This symptomatic change would show in a new mind type behavior, as right action by a few guided ones.

My place was now glowing in an otherworldly golden rainbow light. I was able now to beam back light to the ships. Can timeless thought create thoughtless time? Is thoughtless time equal to timeless thought? These questions are on my mind as I look in silence at the astrological charts of the decades ahead.

As we read these words and study what incoming messages may arise to form thought process in us, we also learn how to observe our thoughts. And in order to achieve this we have to end thought-as-time. When we follow each thought that comes up to its very end, like a trail of footsteps in the snow, we will notice that at the end of that trail of one thought there will be silence. In this space of tranquility resides the Zen of star magic.

As I was thinking this, I saw a huge white light appear in the night sky. Most people think all the time. One thought chases the next, like a hive of mad bees flying around, but not a single of these thoughts is usually thought through and observed or followed to its conclusion.

When we begin what I term the Zen of knowing the self, which is an experiment worthy of our pursuit, we will realize that the process of thinking slows down considerably. While this happens, we are able to isolate one thought from the next, just as a movie director would cut one frame from the next in a scene. The mind then stops the endless chase and the endless meanderings.

For some strange reason the old brain escapes from one frame of thought to the next, like a badly cut movie, while leaving each impulse un-thought, unfinished, and unsynchronized.

The new mind will be able to isolate one thought one by one, in silence, and in that timeless space of utter serenity – which is much more than some mindless mantra repeating meditation – the future mind will find a clean and clear-cut solution, and from there new paths will emerge, where no footsteps in the snow obscure the pathless view.

The emanation of that type of clean thought is sheer magic, minus the man-made memory of confusion. It means in slowed down and almost thoughtless time-space we can manifest a new and clearer reality.

The Shining Ones are none other than the light beings from the lineage of the timeless Dragon Elves. Through the light vortex, they see ahead and have assisted mankind over many millennia in very discreet ways. Certain sages and wise people, who became known on earth, were from the lineage of the Shining Ones. They are the source messengers of Cosmic wisdom.

The golden orbs of the rainbow dragons are mysteriously appearing on peoples photographs, as we edge into the strange expanse of the star gate portal of light. These orbs are real. They are the first sign of contact from the invisible worlds that exist all around us.

Many more strange things will happen in the years ahead. Humans will receive signs from orbs, signs from nature and through the magnificent crop circles. But, more important is what we do with those signs. The mystery has to become a living dynamic within our own heart and mind. For that to happen we have to be the mystery and live it.

The Master of Light is a time traveller creating the world we will live in. We – from a time yet to be – are the guardian race protecting planet Earth. The Cosmic architects of our evolution look back into the eyes of the past, forming our future in the now. There is hope for us, for man and woman, but there is none for this society. The future is born of instant comprehension – communion between two people who are on the same level at one time.

Way of The Magician

"Here is a test to find whether your mission on earth is finished: If you are alive, it isn't."

~ Richard Bach, Illusions
The Adventures of a Reluctant Messiah

The choices ahead are these: If we continue to live as we have in the past few hundred years, we will be on a downward spiral with buzz words and agendas such as – credit cards, debts, data bases, legal issues, medical problems, cell phones, videos, mental pollution, spiritual insanity, digital insanity, highway fast lanes, GPS, face and iris recognition cameras, thumb print ID's, presence detectors, security, accountability via nanotech flu shots, Orwell's *1984* visions, the continuous state of high alert mega emergency and global homeland security, collar controls, and your homes becoming your prisons.

The trigger events will then be computer hacking, global net worms, high tech viruses, cinema and mall bombings, random assassination of the mega power elite, bio-chem events, smallpox infested agents, and the one big scalar energy or nuclear event. The upside of this downward spiral scenario would be: no need for credit cards and debt, dishonesty will become obsolete – no more need to "tell the truth," few unilaterally dysfunctional relationships, reduced time spent on cheating and lying, less sales, less people, reduced retail activity, no real estate... a water world.

Downside of the scenario would be the total, final and comprehensive economic meltdown of a million times the proportion of 9-11, complete system overload, triaging of logistical and medical infra-structures, racial and socio-economic profiling, sky-rocketing addictive behavior, drug abuse, medicine abuse, human abuse, mental illness, mass hysteria, an equal number of prison guards and inmates, more police and army than humans, underclass riots, and global mayhem and chaos. World War IV will be fought with sticks and stones, as Einstein correctly pointed out.

In deciding which way to go, back to the future or forward to cleaning up the planet, we must face the crumbling of the world-wide economic financial house of cards in order to open our view to ancient wisdom and sacred knowing. We must stop "borrowing" from the future, today versus tomorrow, and live in the *Now*, while tomorrow never dies – as the case may be.

We will be forced to address issues of over-population and imbalanced growth, the anti-Darwinian triage of high medical cost to fight the addictive mass populace, the senior euthanasia issue, abusive child education leading to competitive and anti-social behavior instead of playful cooperation and ultimately the understanding of the spiritual life force, i.e., right of life vs. freedom of choice, drilling the planet to death for petroleum instead of looking for energy in outer space, the epidemic disease control, cancer, aging, and all the things we discussed so far.

The "Way of the Magician" implies the implementation of a constructive plan. One that looks like this:

The Western world would reduce its over-consumption and competition-driven behavior. Schools would teach children love and cooperation; no competitive sports allowed. We would understand that we cannot over-populate and abuse this planet's finite resources. War and conflict, drama and disorder would be archaic notions, spiritually repugnant, and obsolete. Violent video games and negative movies or destructive music would be noises of the past, which very few would choose to consume.

Recreational drugs would become legal yet only consumed by the dwindling few outcasts on their way out; or integrated into a sane humanity. People would embrace life rather than run away from reality. Most people would telecommute and home school their children, thus making traffic and mental pollution a thing of the past.

People would live a productive life and do meaningful things to contribute to a cooperative spirit, living well past 120 years of age. Cancer and diseases would disappear. Local energy sources would be free, and there would be no hunger or starvation, no more homeless people. Medical care will have diminished. The world's population would be heading toward a sustainable three billion.

Biological, chemical and nuclear arsenals would be brought under the unified control of a world management authority, and deactivated in a shrine of peace. Advanced energy devices will be used only in an emergency situation, to protect our planet from a meteor or comet impact. All armies would be reduced to nil and prisons would cease to exist, with the exception of mental asylums for those who ran the world until now.

The giant metropolis conglomerations of millions of people such as New York, Los Angeles, Tokyo, London,

Paris, etc., would be decentralized to small and manageable communities. Giant corporate conglomerates of economic super power would be split up into local service providers and healthy, smaller companies.

Crime would be minimal; greater equal opportunity would exist for all, and the few remaining offenders would be treated educationally for anti-social behavior, rather than thrown together in compounds of millions of humans guarded by tens of thousands of guards to produce more mental illness.

Intellectual property notions of the past and copyright ownership would give way to financial support for scientists, mystics, and magicians of world repute. Authors, artists, creators, inventors, engineers, and innovative thinkers would be celebrated and honoured like sport celebrities, movie and rock stars in the past.

The focus would be on playing, meditating, learning, teaching and working together in a non-competitive way. Harm none would become a universal law and new standard of faith. Astrology would be the sacrosanct and final gateway to the star gate.

The web would become a democratically free and accessible global Omni-net or digital village. Virtual reality would manifest as rural living and healthy human behavior. Media and television consumption would be minimal, excluding educational interaction promoting all of the above.

Mindless mental opiate and sedatives for crowd control such as religions would be transformed to a healthy, functional and sane forum of conversation and discussion that would be conducted in neighborhoods and salons, or at home and in fine inexpensive restaurants.

The idea of property or real estate, and time sharing would be obsolete. We understand that we are tenants of time, living on borrowed time, and that our abode is this blue planet which we all help to manage, share and protect. We will look with wonder at small things, cultivate flowers, play with animals, and become one with nature. We will sail and live near the oceans, where life is formed.

We will develop meaningful and longer lasting relationships both privately and at work. And at play, we will contribute to the education of beautiful children who will carry the torch of a galactic belief system based on star magic into a not too distant future.

The new millennium beauty will be everywhere the emphasis being on art, literature, poetry, music and constructive behavior. We will all embrace a new way to align our passions in the pursuit of a new global ethic and aesthetics, while respecting each other's cultures and past roots.

People and races will mix to become one. We are the ones we have been waiting for. No messiah is expected. No heads of states are needed. No governments will exist. We will work and play together as a universal team of humans, devoted to the quest for the absolute. We are our own *messiah*.

On our journey toward the absolute, we will discover the secret formula of life, the cosmic seeds of wisdom, energy healing, remote viewing, quantum mechanic's action at a distance observation, zero point energy, homeopathy, synchronicity, the secrets of the shamans, migration of peoples, nuclear and biological transmutations, and the way to an Atlantean intergalactic society of inter-planetary species. We will become consciously aware, and acutely conscious.

We will ultimately close the giant gaps of knowing still existent by going back to the principles of inter-disciplinary learning. We will educate renaissance humans for a new century and millennium. We will accept what Dr. Robert Solomon of the Intellergy Institute calls multi-dimensional thinking, or what astronaut Dr. Ed Mitchell calls *The Way Of The Explorer*; and what I term the *Zen of Stars*. We will accept truth-based concepts of reality as opposed to myth-based cults and practices such as religions.

We will understand that – to win without fighting is best, as revealed by, Sun Tzu, the ancient Chinese philosopher and author of, *The Art of War*. We will embrace the *Art of Zen*. The main code to manifesting this art consists simply in an old adage: "Harm None!"

"As soon as we realize the disparity between how we perceive things and how things and events actually exist, it prompts us to see through the deception, illusions, and misconceptions of this fundamental ignorance. This allows us, eventually, to release our minds from the influence of ignorance, and from the grip of conceptual thought processes. And this in turn makes it possible for the nature of mind to be released from the influence of negative emotions and delusions, and so attain true cessation of suffering." – His Holiness the Dalai Lama

*If we can predict the outcome,
can we change it?*

Total Change

"All authority of any kind, especially in the field of thought and understanding, is destructive and evil. Leaders destroy the followers and followers destroy the leaders. You have to be your own teacher and your own disciple. You have to question everything that man has accepted as valuable, as necessary."

~ J. Krishnamurti,
Freedom From The Known

In the past we have built much imagery, created icons as a comfort zone and psychological fence of security. These icons are spiritual, religious, political, economic or personal, and we built them because we committed the error of believing in something, or someone. Belief is based on fear. Fear is based in a mal-functioning notion of psychological time. Awareness without thought is based on truth. Truth is always new, never the known.

Belief is based on dependence, and it is born out of the idea that we are not good enough, or that we are not capable. People no longer feel sovereign, responsible for self and capable of being that which they *believe* in. They make an outside force responsible for their spiritual development or strength, and they follow that force. The *belief* in the outer force is a weakness of the inner force.

Zones of comfort, which are fenced-in unrealities of a deceptive psychological nature, are evil. They manifest their addiction to beliefs through symbols, ideas,

433

ideologies, policies, religions, convictions, certainties, creeds, unproven axioms and beliefs or belief systems encompassing the mundane as well as the spiritual. All of it is total nonsense or collateral damage of our moral evolution, because the movements are utterly senseless, and without any real direction. They do not exist, yet they led us to failure. It was all an illusion, as we said at the outset of this book. It is time to get off the merry-go-round.

Belief is, in the 21st Century or *Age of Aquarius*, a most cumbersome intellectual burden, which slows or brings to a halt humanity's progress. Attachment to beliefs and ideologies have led to global war, famine, political, social and economic upheavals, destruction of our habitat and general dysfunction on all levels of society because they divide us from each other.

Few people see why it is violent to call oneself an American, Russian, Swiss, Muslim, Christian, Hindu, or a whatever. It is violent because it is divisive. Think of yourself as a human being. Peace is something that does not suit those who feed off the planetary divisions.

Human perception of life is shaped by concepts already established in the collective conditioned mind. The time for truth has arrived, the time of delusion is coming to an abrupt end. Pluto is inching toward the transformation of society, as it does every 250, 500 and 1,000 years, and there is no escaping the transcendental effect of the "Lord of the Psyche" when it reaches the entry into Capricorn in 2008.

The Sagittarius key word is *I see*. The sign of the archer takes no nonsense and is always right and truthful. This means by 2008, Pluto will have cleaned up religions to their bare bones, integrating Krishna, Buddha and

Christ into what is truthful, i.e., scientific and spiritually correct. Capricorn stands for *I use*; thus, Pluto in Capricorn until 2022 means what is not useful must go. It is that simple.

The new human child is an ancient *soul carrier*. The uniqueness of a person does not manifest in the superficial, it manifests as freedom from the known. The individual is a cosmic being. Freedom is observation without fear. Freedom has no motives; which is not at the end of the human evolution but is the first step. Total change is not easy, is not comfortable or desired. But the uneasiness of the genetic-light DNA revolution, led by the *soul's invitation*, is the necessary turn of events we have to manifest in ourselves.

Krishnamurti said that, freedom is found in choiceless awareness, and that thought is time born of experience and knowledge. Thought is not intelligence. If you observe carefully you can see how man is a slave to the past. When man becomes aware of the movement of his own thoughts, he will see the division between the observer and the observed. When he was in his mid-thirties, in 1929, Krishnamurti dissolved the "Order of the Star" with these famous five words: "Truth is a pathless land."

He ended his chilling speech with the following: *"Total negation is the essence of the positive. When there is negation of all those things that thought has brought about psychologically, only then is there love, which is compassion and intelligence."*

In other words, we can pretend to ignore certain things, yet there are situations we must pay attention to, because our silence would make them appear as if they are right, when they are totally wrong.

The great poet-author, Robert Anton Wilson, said: *"My own opinion is that belief is the death of intelligence. As soon as one believes a doctrine of any sort, or assumes certitude, one stops thinking about that aspect of existence."* He is not saying that people do not have a right to their beliefs. He just says he doesn't believe in anything, and neither do I.

If people want to surround themselves with beliefs and thus live in fear, then by all means, let them. Let them live and die with their beliefs and their fears. But for those who do not want to close the door, and who wish to inquire, look and understand... life takes on a completely different meaning when observation is free of belief.

The bright minds of the 20th Century, such as Immanuel Velikovsky, Tesla, Reich and many others, communicated and applied their knowledge because they felt it necessary to share their research, discoveries and gifts with humanity. I personally, would have had an easier life if I had kept what I know to myself and just tried to fit in. Something I found impossible to do. I feel I would not be following my destiny if I did not speak out and tell you what I know... and explain to you what will happen in the future, and show you what you can do about it.

The world has now ended its *childhood*. The earth is at *Childhood's End*. Now comes the awakening and the middle age of planet Earth.

The Master of Light had stated to me things of great simplicity: The destroyers of the earth will be the ones facing their own terminal demise. Then they will understand what it means when we say, *"What goes around comes around."* I will be on my way home by then.

But I am on the side of the future, because I am the one who came back from that future so I could explain it to you. I came back so that we would end the games of division. It is up to you to decide whether you want to ascend into the future or perish in the past.

All you have to determine now is your stance. I have shown you seemingly complex things through a mirror of seeing in a civilized manner. I have used simple words, minus the unnecessary occult mystery school for the retarded elites who thought to keep the truth secret from you. This is the time to emphasize solutions, and I hope I did.

The key message is: you must not give your power away to anything or anyone, not to a belief, not to a religion, not to a philosophy, not to a person, nor to a government, nor to an alien, and not to any thing, idea or concept. You reclaim your power. That is the message of *The Master of Light*, and therefore it excludes any teachings to follow. It is now the age where the teachings come from within.

No savior, guru, avatar, ascended master, belief system, church or religion, and no institution, or organization, will save you, nor will any belief in any thing help you. No government will care for you. You will organize yourself with friends and neighbors, and choose the best among you to manage yourselves; because you are your own leaders and creators of solutions. Most belief systems were created by alien-parasites to control people. The true new world people will not be managed, they will manage themselves.

We have come to a **time of total change**; not the change claiming to be prophecies of a time yet to come, but the trans-soul transmigration that bears the signature of the creator of all that is. That signature is an ever increasing sound; it is the light-sound, and it will be heard across the whole earth in the transformative years ahead of us. *The Special Ones* will know how to deal with this total change, and I salute them.

The great Indian Chief Seattle declared at the signing of the treaty in 1855:

"When the last Red man shall have perished, and the memory of my tribe shall have become a myth among the white man, these shores will swarm with the invisible dead of my tribe, and when your children's children think themselves alone in the field, the store or the shop, or in the silence of the pathless woods, they will not be alone. At night when the streets of your cities and villages are silent and you think them deserted, they will throng with the returning hosts that once filled them and still love this beautiful land. The white man will never be alone. Let him be just and deal kindly with my people, for the dead are not powerless. Dead – I say? There is no death; only a change of worlds."

A Future Vision

"You must be the change you wish to see in the world."

~ Mahatma Gandhi

I do the work I do using an intense focus on only that and I do not allow my energy and focus be dragged into the world. Each one of us similarly has a light of focus but the world steals the light of most people and uses – or misuses it – for its own ends.

You only have to learn to withdraw your light focus from the many distractions of the world and focus it on your soul purpose, to know what I mean. That purpose can be a number of things, and when you apply it, your focus gives life to your purpose *in this world*. All you see me do in this book is to give that focus.

You may hesitate and doubt that you have a purpose in this life, because many distractions are stealing your focus, your intent, in which case you feel fuzzy, torn apart, unhappy or weak. For example, conflict is one of those thieves of the light focus, which then makes you feel torn apart. But each conflict also needs your cooperation to effect you in this negative way and to distract you from your power.

To activate your own focus and purpose, all you have to do is learn to *see* the actions that steal your power and don't participate, step back, look before you leap, create a gap... and once you apply the art of staying focused you also learn the art of yourself – and apply it.

The vision I share with you on these pages exists in my heart and in my mind. From the heart and mind we manifest our reality. The only viable future will be developed on a small community level, rather than on a large centralized one. It is clear that it was a mistake to allow the larger social infrastructure to dictate community development with a centralized structure of resources. The structure now in place dehumanizes society. It makes all of us dependant on a central government and a central banking system, and the *victims* of a central elite group that rule over all basic and fundamental resources world-wide.

By withdrawing local and community responsibility for environmental issues, the door was opened wide, allowing corporate abuse and placing business interests before health and planetary welfare. Since water and air pollution became the concern of the state, the corporate destruction of the environment followed, under the control of corporate-serving politicians and businessmen. In forming this global-strategy of resource control, it was clear that individual politicians could be more easily manipulated than local populations, especially those who suffer the consequences of drinking polluted water, or breathing polluted air.

I envisage an ecological peace development plan with a program that encompasses all areas of social and cultural development. The aim is to re-establish human sovereignty. Human beings must now create a new future. This means not only cooperation on a personal level; it demands practical and effective compassion as the primary operative for economic, business, housing, food, air and water resources. Ultimately, a network of like-minded spirits will manifest international peace, cooperation and understanding. This in turn will drive an entirely new economy, based on fair trade rather than the present corporate greed for profit.

I intend to create a Trust via our own banking facilities, with *Integrity Resourcing* as the main objective. We will develop models that can be used world wide, but not within large cities, whose very structure and maintenance are against sovereign human development. These models will be for smaller communities and poorer areas of human society, where food is produced, but where the investment is lacking. The first model is envisioned as follows:

St.Clair's Cosmic Contact Center – or the *City of Gold*, is my vision of the future. Groups of people will create small self-sustaining eco-villages all around the globe. The intention is to create a haven for people of all races. It will be founded as a financially viable and economically secure self-sustaining project.

Only a few people will live there, with others spending a few days or months each time, to study and to teach, to research, and to assist my core group in international projects. The *New Earth Project* is designed to blend with the surrounding landscape, giving protection to the shelters and yet harnessing the elements of beauty and aesthetics to bring balance and peace to the setting.

My Retreat brings innovative individuals from all over the world to a place of peace. Guest will carry on discourse to explore and exchange knowledge. It will be a center of earth energy that will benefit those coming there with the intention to serve, while the earth changes are taking place. The Retreat will be built with great care in order to establish a balanced setting for launching the *contact center*. The essential core is that we carry seeds within us, and it is time to plant those seeds and create a new future for all mankind.

The Contact Center is a space to connect with yourself, to connect with your vision, to contact the ancestors and the invisible worlds, which will function in co-operation with the retreat. All who are interested will be welcome to make use of the knowledge resources at the center. Here I will have assembled the leading library of the world. The basis for all activities associated with the *cosmic survival center* will be an awareness of, co-operation with, and respect for the earth and the other planets.

St.Clair's Earth Retreat will be established with emphasis on navigation by the stars. An earth survival school will teach people of all ages how to connect and live with the wilderness – spiritually as well as physically. These skills will be taught together with an awareness that the earth is a living being who provides us with everything we need, and who should receive our thanks as part of our ongoing relationship and friendship.

The Cosmology Center will offer the opportunity for personal study and private consultation, and the chance for those who are interested, to develop an understanding of the skills of an astrologer. Apart from individual consultations, the participants will learn how to better understand the planetary forces. I will show how to empower, meet and address circumstances, rather than be their victim. Stars will become a living dynamic in the lives of the participants. It will become clear that each of us has a powerful relationship with Cosmos that teaches us peace, joy and stability in our daily lives.

The Psychic Recall Center will teach participants how to use these skills in all areas of life, while sailing, climbing, diving, learning survival skills, healing, while studying Astrology, setting up a business or a strategic economy and in daily life. Psychic Recall will be used as a tool to connect with the inner mind, and participants will learn how to make use of this tool in everyday life, at home or at work. They will learn to see beyond the patterns and resolve to change them. We must know how to use *the sight*. The core group of sages, seers and shamans assisting me will find new technologies suited to develop a peaceful world. Participants will learn to direct and use the Pranic or Chi energy-shell force, to establish and maintain their own energy balance. They will learn how to develop an intact energy shell as a powerful support to the healing of the body. The psychic force field is the self-defense shield.

The Energy Engineering Center will embody practical techniques. These will include the use of plants and herbs to assist in healing and strengthen the physical body. We will also learn how to use the mind as a tool for self-healing by talking with the body as a way of re-establishing necessary self-contact. The direct healers will be the mountains, plants and trees, concerned with empowering participants to find solutions and take responsibility for their own health. The body-mind is the key to physical and spiritual health.

The Meditation Center will be a sacred place where people can receive peace of heart, peace of mind, and ultimate serenity under the stars. It will be carefully positioned to balance with the earth energies and resonate with human frequencies and energy systems. This concentration of silence will be the bridge to Cosmic light contact. It will be an earth and star bridge, connecting past, present and future as a unifying life force. This sacred space is the core of everything that will take place within the survival, teaching and contact center.

The Organic Hemp Center will provide necessary food for those residing at and visiting the retreat and contact center. It will also offer a means for visitors to have a practical experience in working with the planet in a balanced way. The Hemp farm will enable people of all walks of life to experience and take part in the production of their own food. They will learn to understand that our relationship with the plant kingdom is essential in creating a healthy human being as well as a healthy environment. This food production center will create an economic base for the whole community. The key plant will be Hemp (the low THC variety), which can be used in hundreds of ways, clothing being one of the many uses of this amazing plant. *Buddha was reputed to have eaten one hemp seed per day on his six year fast before attaining enlightenment.*

St.Clair's Peace Center will coordinate peace projects world-wide. It will be the magi's center for co-operation, crossing all social, religious and cultural barriers in establishing a vibrant, viable relationship among people of all ages and backgrounds. The peace center will be the foundation for discussions and lectures by experts in the Sciences of Astrology, Alchemy, Cosmology, Healing, Economics, Native Cultures, Literature, Music, the Arts, and all forms of cross-cultural sharing of ancient wisdom and new knowing. I will connect a network of like-minded centers around the world to produce synergies and concrete accomplishments toward peace and understanding, during and after the earth changes.

I was told that this may well be the most ambitious part of the project. Yet without it, none of the other components makes sense. I anticipate centers like this to flourish throughout the world and become a vast network of inspired and inspiring focal points of light, joy and natural spiritual growth.

New Communication Technologies will be an ongoing synergistic interface with my clients in the areas of self-administration, marketing and public relations skills, and in the establishment of independent public and private communication services. This will include the development of new economies and research into the art of telepathy. We will have our own telecom facilities and develop new computer languages. The center will also feature writers and authors for special events.

The Orgone Art Center will be concerned with creative skills of merging alchemy, science and art into one focus. Physical creations have a magnetic and energetic signature, which interacts with the surrounding magnetic field of the earth. Emphasis will be on creating art that can be used. One example of this is the beautiful Anasazi pottery which is both functional art and design. The discoveries of Nikola Tesla and Wilhelm Reich are also art forms and can be applied to our world as functional art.

Water World – Water is the single most important substance for life on our planet. All living beings are made

445

up of large amounts of water. We humans consist of over 70% water. It is therefore vital to have healthy *living* water to sustain us.

Clean water is the most precious and essential element on this planet. We need water for our sustenance even more than we need food. But the water has to be clean. Polluted or lifeless water creates a welcome environment for disease. Communities must develop their own alternatives to centralized wastewater treatment. It is senseless to send wastewater through leaking pipes to a central water treatment plant, when the process of treatment can be far more efficient and economical. Living Technologies, for example, have developed an efficient ecological waste water treatment system that can be used locally and perform to the highest standards. A *Living Technology Machine* is in use at Findhorn, Scotland.

As the western world's redundant centuries-old infrastructure collapses and dies, the state-of-the-art development of St.Clair's – *Center for Cosmic Survival* will concern itself with cutting-edge natural living technology of the future. It will seek ways to promote ecological engineering as an efficient and superior alternative to our current water supply and filtering systems. My Center designs include sustainable solutions and the research for free energy.

I have advised some of the leading scientific minds searching to develop solutions. In private symposium meetings I show my clientele how we must prepare technically as well as emotionally, spiritually as well as mentally.

The Future Is Now

"We shall no longer hang on to the tails of public opinion or to a non-existent authority on matters utterly unknown and strange. We shall gradually become experts ourselves in the mastery of the knowledge of the Future."

~ Wilhelm Reich

Someone once told me that I am a mirror to others... that they could see themselves in me. This makes for an unusual childhood as we discussed early on in this book. We are all of Atum, of the stars and of the earth both, or else we would not be here now.

Relationship with life shows you who you are, so why do people turn to a spiritual authority to find out what life is about? Every one of us enters the physical earth experiment with spiritual eyes. Obviously, most have forgotten the true birth experience. Why ask another source who you are or why you are here? The reality is that each of us has the inner capacity to remember ourselves, if we look hard enough.

The invisible world of ancient beings has patiently shown me how to take command of my own life. I have received very few direct answers to my questions. It happens from time to time, but it is rare. The way the answers come through the world of spirit is as a catalyst to discovering the answers for myself. Events will often happen, that trigger an inner knowing or a realization. The truth revealed to me is that I am both the teacher and pupil, or the master and the apprentice.

When you are learning, time is irrelevant. All that matters when you are learning (teaching yourself) is the inner link, or the inner awareness. You can call that your guidance-navigator. It is like a golden thread of awareness. The light of awareness listens and learns, turning each experience into a wordless dialogue with a *Master of Light*. At the centre of your life is a pure flame of light. All experiences are returned to that flame. The entire experience of incarnating across time and space is brought to that flame, the source of one's own being.

I am not sure what happened to bring this awareness to me. It just happened. It took place one day before a full moon and lasted for three days. It was strongest during the full moon and continued into the next day. In the morning, I woke up and was strongly aware that a powerful spirit was resonating together with my physical form. I could feel how my spirit was sticking to my body. An awareness of restriction and lack of freedom made my physical-material being feel very uncomfortable.

With all those planets, galaxies, super novas, and central suns out there, it felt unfair that my eternal spirit was restricted to this body. The feeling and the awareness became almost painful to me, as it raised all kinds of questions. How can spirit remain inside the physical experience? The feeling of restriction grew even stronger. Are we not seeking freedom? It did not make sense that spirit forms would seek freedom by becoming trapped within a restricted physical reality.

Nevertheless, it is freedom of some kind to be a guest on this earth planet, and who wants to live forever? I began to understand that without the restriction the spirit being would roam through cosmos with no desire to return to local space-time. Interacting with physical time we become the alchemists of our own life experience. For a reason!

Experiences such as these are fascinating. They show us who we are, if we are willing to look at the experience as a mirror that shows us our own inner reality. The actual experience, whether pleasant or unpleasant is not important. What is important is what the experience teaches us. The reality inside of us is the Zen garden of our own existence, and of our own making. It is the reality of experiencing this world, and this *is* learning. As the inner reality looks into the mirror of life, an eternal presence is reflected into and from that mirror. The key to freedom lies in our ability to look straight into the mirror by seeing who we are. The vastness being reflected in the limitation of the form we have chosen for ourselves.

The free spirit contained within the restriction of form is experiencing something of the highest level of being. The desire to escape the restriction of physical reality divides us. This division is the desire to experience another world, yet this world is not something we have to escape from. When humans talk about freedom, it is usually related to escape... the moving away from something to achieve freedom. To create any reality, it is necessary to compress that which is limitless. If there are no borders, there are no experiences.

The feeling of my spirit sticking to the physical body and being restricted by the world of flesh was, at first, unbearable. I had achieved something I had searched to understand since entering physical time-space. Through the light of the full moon, I was perfectly aware that I was spirit in human form. I could equally feel the presence of my spirit-being and what it means to be free. At the same time, I could feel the physical restriction and what it means to be restricted.

The strange symbiosis between the two was shown to me through two equal and opposing forces. The free flying spirit feels powerfully drawn to experience the oneness of being. The physical instrument seeks to remain grounded in containment in order to continue its existence. The symbiosis is hidden from us, because it would interfere with our physical existence. At the same time, an awareness of its existence is necessary in order to understand the nature of this reality. Peace of mind and peace of the heart is achieved when the human fully enters the human experience.

Ultimately, we destroy our own freedom through fighting the experience of being restricted to our physical form. As I observed the two forces creating my physical manifestation on this earth, it became clear that the beauty of this unification is as awesome as the universe itself. Unite the two and enjoy your journey through this life. The unity of spirit and matter magnetize reality.

Humanity and this world are in the throws of a cosmic quantum shift of vibration and consciousness transmutation. Things will all of a sudden totally morph into something completely new, and then what? We are standing today on the brink of the cusp of an age. There are those with prophetic vision and insight – or psychic recall – who have incarnated on the planet today as harbingers of this impending transition. Some of them, kids among them now, are seers.

Such visionaries are tireless seekers of truth, their

souls are dedicated to serving that truth and their hearts are devoted with an uncompromising focus upon the plans of the divine. Their true purpose for being in this world during these times is to be consecrated to serving as clear and purified vessels through which Cosmos may pour forth its energies of knowing. It is the *Zen of Magic*.

Part of what they have come here to do is to peel away the facade from the lies, distortions, and corruptions that compromise humanity and debase the body of planet Earth. Another purpose is to allow the light of truth to shine through them upon those shadowy and long denied aspects of our world-wide culture, thus exposing the lie upon which the existing paradigm is based.

It is apparent that few among the mass of humanity are prepared – at least as of yet – to see what is being revealed by such visionary insight. Most people confronted with these prophetic insights not only feel challenged by them, but also deny and repudiate them, but this will also pass.

As a species, and especially as those who have been born and bred in the first world – the so called civilized world – we have been deeply conditioned to think, feel, speak, and act according to certain particular selfish and fear-based patterns of thought, emotion, word, and deed. This collective patterning has contributed to the prevailing paradigm on the planet that is responsible for the systematic degradation, decimation, and defilement of life, as we know it. What we are living through now is the end of the dominator culture before we move to a partnership culture.

Those seers, who have come to presage what is about to unfold, are also here to untangle the web of lies and habitual patterns which are based upon fears and desires of and for the self. As the fabric covering the lie is permeated, and as the light of truth shines upon that which is shrouded beneath the veil, ultimately the lie itself will be seared by that light and transmuted by its fire. It is through such revelation that humanity may ultimately become purged and cleansed of the dis-ease by which it is currently possessed.

Mirrored layers of reality have always fascinated me. The mirror of seeing, of Galadriel, that Frodo steps toward, shows more than one... *wants to see*. It is somewhat inscribed in my birth chart that I would deal with several levels of communication, both in the actual foreign languages I speak, in telepathy, and in codes. These are color, tone, archetypes, and the *Icons of Destiny*. It was also inscribed into my chart of birth that star knowledge was handed to me before I came here. I have shown some of it in this book. Handle it wisely please.

To be frank, not many people were able to understand me. I seemed removed from them, just as I am today. Only later will people realize I came to meet and greet them from the future, back into their time. Although I don't really know more today than I knew when I was a kid – such is the way of the mirror.

The way of the magician, the master of light, psychic recall, inner sensing and future vision shatter the matrix illusion – by using *the sight*, the angle of reality woven into dreams, dreams being the rainbow dragons of our reality or golden orbs of awareness.

Our dreams weave reality and our reality weaves into our dreams. I came one night to a place I know so well, and yet within this physical reality it makes no sense to say this. Somewhere in the mountains close to the sea, there is a community of people related to us, yet through this time-space they remain as far out of reach as Shambhala. They may talk to us in our dreams, perhaps to bring the two realities together as one.

In the place that I visit when I am asleep, colors have a different depth. Dreams seem closer to reality. The mountains can be felt as a living reality through the psyche. Emotions are balanced because they are connected in a healthy way to the web of life and not simply to the desires of the self. Texture is sensed through the mind by the mind, and is not merely a physical awareness. As hard as it is to live through this present time, some of us are connected to the future in an advanced spiritual way.

The power we use to balance this present twist in space-time is the creation light we share as our collective future. We are ourselves in the future looking back. The moments we share now are the building blocks of that new world. It is not through chance that we visit those sacred places in our dreams, sharing their realities and bringing that reality back to the earth with open eyes in the morning light.

Exploring the future is what I like to do most. By default of my unusual character my life is not *normal*, well, either I am not normal or society is not normal. I prefer to think that I am out of the norm. It gives me more freedom to shape shift and to observe the world in ever new ways – while exploring the future.

Time is not a notion people have a good grip on because it is a multi-layered reality, not to mention a subtle illusion. A door had opened in time early on for me, and it remained open forever. Since then, the two forces –

time and its doorways – have accompanied me through this adventurous and unusual life. This is in essence what *The Master of Light* had shown me. You are your own future. It is inside of you. Access it!

We are beings of joy and of light, although some of us have become rather sorrowful believing in the divisions and conflicts of service to self. By nature and in my approach to life, I am an explorer. My birth planets suggest that I must study life to appreciate it. I enjoy the company of people and I value the bond of friendship. The spirit of the invisible world taught me how to stalk the unknown, to seek out uncharted spiritual territory, to remain hidden in plain sight, while I watch, see, listen and observe...to discover new places, and then to take this information back to the people moving towards this reality.

I explore the dangers and find solutions. I see the risks and the opportunity inherent, and I do not mind experiencing the hardships and find a way through them with the least effort. When a community has to survive, it is vitally important that they can apply the least effort to attain the most balanced sustenance, psychic as well as physical. This planet Earth-ship is a community attempting to come back into balance. Some advance knowledge would be helpful to accomplish this mission.

The landscape most overlooked by people is that of the psyche or the inner Zen garden of life. This indeed is strange because without the psyche, not only does nothing exist, but there is no journey through life. The physical body alone cannot make its way through life. Yet, people remain attached to the physical material world; although at death it is the physical body that remains unmoving while the spirit departs to unknown lands. That is when the soul journeys on. In that sense we might as well get used to seeing the mirror of the soul.

I have the highest respect for my spirit knowing. Spirit is truly the master of reality. I do not know how the spirit manages to manifest the future in the now, but that is exactly what is going on. Maximum effect by minimum but focused energy, just as my Zen master type instructor in the Swiss army taught me. He was a colonel who knew some of these things. He could read people like a book, and adjust reality accordingly.

Most people will, mentally speaking, not be able to handle this shifting situation or phenomenon. Some will go quite insane, and many will simply and quietly pass away when this happens. When that development begins at a mass level, you want to be somewhere safe and sound.

The Master of Light, who showed me the Zen of knowing at my castle, impressed on me that we will have to imagine another reality in order to create it. Create it we must at some stage. This entails releasing the creative spirit of the mind, shifting away from the thought based limitations and into a higher Cosmic vibration.

The upcoming eclipse series activate changing events for many years to come, well into 2020 and beyond. It is a sign that we are coming closer to the period of transformation. The trans-human souls who can communicate this are going to be our friends. The time is now approaching. It is happening. The crossing of the time zones is very near. In this time yet to come, it is vital that a minimum number of people can make the psychic shift before the time zone crossing happens physically – when the invisible world becomes visible.

The Empty Hour Glass

"Migres, migres de Geneve trestous
Saturne d'or en fer se changera,
Le contre RAYPOZ exterminera tous,
Avant l'a ruent le ciel signes fera."

"Leave, leave Geneva all of you,
Saturn of gold will change into iron,
The counter RAYPOZ will exterminate all,
Before the advent the sky will make signs."

~ Nostradamus (quatrain IX.44)

The sky had shown me many signs. Extra-terrestrial space ships, blinking lights, celestial omens, astrological alignments, green language and I are close and old friends. Yet this is the one prophecy – IX/44 – of the French seer that no one dares to decode.

The economy (Saturn) would turn from banking (gold) to something simpler – iron, barter. *RAYPOZ* or *ZOPYAR*, God of winds, or the use (Saturn) of the ion beam, opens many possibilities to interpret this key. Does he see *CERN* (European Centre for Nuclear Research) and its accelerator going South?

Who would want to oppose *RAYPOZ*? Which planet shall Saturn face down? Is it a known or unknown planet? Saturn is also known as the *Time Lord*, as well as the bringer of form. The prophecy talks of the changing of the ages, of the change of worlds and maybe even of the pole shift. The time frame is vague...

The eclipse was masterful, in fact picture-perfect. The rainbow circle glistened above the misty mountains of Zen magic. This time the *Master of The Light* stood with me in the eight hundred year old hall of the prophets at Chillon castle. There was this immense vastness of the sacred emanating from his presence, filling the room. He had taught me to question everything and to seek right action. This was the final initiation into seeing. After this all else would seem trivial.

Together we looked at Joshua and other master prophets painted on a ceiling by the master designers of the Templars in the late 12th Century. The colors merged with the rainbow lights cast through the high Gothic windows. Aquarius, the water bearer, was painted into the apex of that crypt, faded somewhat, but still pouring knowledge from his urn. What was the knowledge about? Human evolutionary process will be accelerated now and a super consciousness will emerge in mankind.

I wanted to ask him many more questions. The answers to those questions came from the core of his being. I wanted to know, where is the absolute? And who are you? There was a very long silence. He would always go silent at first, so that in the space between thoughts and words a deeper mind to mind contact was established. *"Your unconditioned mind has to have the freedom to find out."*

The man from the future was looking through the time portal that he and I had created eons ago. His look pointed me to a boat floating close by on the lake. To the unsuspecting onlooker we were merely gazing into the lake of Geneva through the Gothic windows, or watching these timeless mural frescoes, maybe we were meditating

– or something of that sort. No one seeing the two of us together among the foreign visitors would have been able to guess what was going on during the eclipse of the light sound carrier.

However, in its true essence we were seeing the future from a time yet to be, by gazing through the mirror of seeing which the lake had now become to me. We were in the process of generating energy that never dissipates. It was to be magic applied in action, the action of the enlightened peaceful warrior, which he was introducing me to, birthed as we walked, to explain to me the end of dominance and the notion of renouncing survival. He showed me what he saw, through a reality window some would have mistaken for a dream.

They say sages never dream. We are always effortlessly aware and alert. Reality weaves itself through our dreams, just as our psychic dreams network their way into our daylight reality. The time portal of Chillon and its window into timelessness opened on what I know now to be the edge of reality. The seeing once again became its own action. The only way to predict the future accurately is to create it – by coming through it in the way we see it fit together... by living it.

What I saw was unusual to say the least. A highly efficient and well-trained group of men and women entered my field of vision. They were dressed in dark grey engineering type suits, the kind you see on a space station. The group of people made a highly cohesive and disciplined impression on me. One could say they were almost militaristic in the style of their appearance, and yet they were clearly mentally far more developed and much more advanced on a human level.

There was an intense silence behind the group, as a brightly-lit presence shone through a glass pyramid below the surface of the lake, and a white room – hidden from the strong light that filled the main area they had entered – opened to reveal their world to me and make my world theirs. Their world seemed to function very precisely, without noise, and most peacefully. The instruments on display were *aware machines*.

The leader of the group of eleven came forward as if to welcome me into his sanctuary of sanity. His demeanor showed a high intelligence. From the way he moved it was clear that his training was exact. His behavior was beyond efficient, energetically engineered to be meticulous, as he transmitted his thoughts to me with precision and with an awareness that translated into an almost super-human sureness. His sky blue eyes betrayed a glimmer of gentleness; it was a look beyond hope. Some would call this compassion.

The five men and the five women who worked under the command of this time lord were obviously used to being together. A sense of loyalty to the mission at hand, and a strong unspoken understanding that goes beyond telepathy and into other realms emanated from this core group of advanced trans-human souls. If I had observed them on a dark night standing near the intersection of a lonely highway, I would most likely have assumed that they were not of this world.

Truth is sometimes stranger than fiction. These people were, indeed, not of this world, and yet in it. Strangeness must co-exist with strangeness in the times we now live in, and so be able to be welcomed by what seems to be a group of *visitors* from across the bridge of time. Even more strangely, what if it is we who are visiting them, in their space time? What if we are the *other world* visitors? How would you then act entering *their world*?

The Master of Light had educated me over something like forty years in this lifetime. Maybe he and I had known each other for a span of *time* that would seem unrealistic to most, and yet the knowledge he and I shared was not of this world and not of this time. He had prepared me for this meeting of the minds, and he had taught me how to see. An infinite energy operated through him.

I knew this group of time lords had contacted me because they wanted me to assist them in their final work of art. What for? Why now, and how? From the first contact, I had learned I could send my thoughts to them and they would answer. They were inclined to answer me in wordless ways, sometimes signs and other times direct. I could understand their varied language and ways of communicating. In fact, understanding them opened up new areas in my mind. You have to use the tools you are given, or they won't work for you.

Elements of my life interconnected like the crossing waves at the sea shore. There was no beginning, and no end to this teaching.

* * *

I had asked *The Master of Light* about the most enigmatic alchemist, the man known as Fulcanelli whose identity is mysterious and unknown. Fulcanelli published, in 1926, his masterpiece, *Le Mystère des Cathedrales*, as a small limited edition of 300 copies. This unusual book is still the talk of European *occultism*, and alchemists. In it he wrote:

"The age of iron has no other seal than that of Death. Its hieroglyph is the skeleton, bearing the attributes of Saturn: the empty hourglass, symbol of time run out..."

The Master of Light smiled when I asked him about Saturn turning from gold to iron. The so-called: "Golden Dawn" occult will return now as hard core iron science. We reach the ending of time as Krishnamurti and David Bohm discussed in the 20th Century.

The time lord type astrology has the building blocks of time and energy encoded in the stars. While Nostradamus left time capsules no one can understand and the *occultists* codified the uncodifiable Nature in cathedral and castle stone walls, we are left today and tomorrow with right action based on our anticipatory guidance. This is why the *High Elves* use light as their language, so that things become more evident.

When the leader of the time lords stepped toward me to show me their science, he too operated with light language and I received an incredible amount of knowledge and holographic levels of information about who they actually are and why they appear here now, in our time zone – while Neptune is in Aquarius – and at this particular passage in earth's history.

The old brain was dead. A new mind had taken shape, one that needed no memory that interacted dynamically with the sacred knowing that is part of the ether, the building blocks of the material 3D world, one that re-invents itself. The purpose of this group of sages was for the good of mankind, and they are here to redeem their own past, by remembering our future. They show us the way through. By learning their secrets I am now able to understand what this existence is about, and how we will change the outcomes.

They were and are what you could describe as shamanic sages, like the Jesuits of space, but not systematically structured, and although they wore no visible insignia on their grey suits they would have left you with the impression of esoteric seers of the higher levels or interplanetary space-time travelers.

They reminded me strangely of my Swiss military days when among higher officers no distinction was necessary. One was among them or one was not. The rest was, literally speaking, silence. The time travelers were careful not to let any hovering darkness make a foothold into their locational presence, when moving through the fields of vibrations. Their mission was a scientific one and had nothing to do with politics and the everyday fears of today's society. These men and women were highly advanced explorers of realities beyond time. In fact, they make the new realities happen.

They had no need for recognition of any kind. I know they had transcended the desire for achievement, accomplished as they were. They were here to complete their mission and yet they were detached from it, and so far ahead, that it felt almost weird to be in their presence. They made me understand that I was one of their own, a precisely designed work of art, an alchemical outcome of their science – Cosmos applied in creation, astrology in motion, a mandala enlightened – if I wished to partake of their adventure.

As I interacted with them through a high level of seeing, I was initiated into their operations. I became aware of the inherent dislike they felt for the cruelty of humans. It struck me that they were not pleased with how we had progressed as a soul group, for failing this *Cosmic test*. They had the coordinates for our new world and it would be a one way trip. I was invited onboard. The childhood's dream was coming true.

Their society was part of an ancient knowing – probably of Tibetan origin and from the star system of Sirius. They had, over millennia of our calculations, reached the utmost level of scientific development of renewable energy, sustained environment and defending themselves in non-violent ways. Their civilization had reached its pinnacle. It functioned smoothly, silently, peacefully and with absolute precision. They had, beyond any doubt, the answers to our questions and they were the solution to our problem.

They told me that they were the descendants of a group on earth who had long ago developed and successfully used magnetic-scalar technology for many purposes. They had learned to manipulate time, space and reality until, as their leader said to me, they had over many lifetimes become who they are today. They were what seemed to me the perfect men and women. They told me that their world is our tomorrow, if we choose to go in that direction.

The men and women of the clans of light were the ultimate product of careful breeding and of precise energetic engineering. Their horoscopes were of such precision as to be stunning mandalas of power, projection holders and hooks to catch my attention, as we looked at their new geometry and at an astrology that was spherical and totally outside of this realm. We talked about star systems looking in from afar, how Cosmos is direct action and how we access this through an inner guidance of our own, or direct knowing.

Their overall life span is ten times the ordinary human lifetime on earth. They had eliminated most of the undesired sickness of their society through a peaceful biological application and selection of birth harmonics. Transforming manipulations of the human gene were applied in their evolutionary journey as a species. They said to me that they are human beings when I asked them where they are from.

For generations they had carefully followed the secret protocols of star knowing – what I now know as the *Zen of Stars* – which entails moving towards some visionary goal by volition and with intent, but with a renewable energy of the mind. They sought for and eventually found together the knowledge of the ancients. They had figured out how we would progress. We would do this by awakening another part of our mind.

I had wanted to know about how we defeat what I term the *dark force*. The explanation I was shown was amazing as we prepared our joint dream of entering enlightenment.

The dark force, that had arrived on this planet Earth eons ago, and began to feed from the lower level human ignorance; is using its own forces, in this case the political and economical leaders of our current world, to plunge mankind into a permanent state of fear.

This state of fear and ignorance has lasted for at least two and a half thousand years. However, by opening the doors to this dark parasitic force, the elites and the dark ones themselves became victims of their short sighted strategy. This proves how inherently unintelligent they are. They are terrified of those who *see* their signature through the presence of the dark force.

The evolution of the human species – in matters both spiritual and physical – progresses only for good when we pass from one level of consciousness in the etheric density of our galaxy and planets to a next or higher level of knowing. This leap of mankind has to be taken as a group, and yet like the Salmon returning to their source in the river, human soul by human soul. Together and yet alone.

It is actually not *mankind* that can progress as such, but rather, it is Cosmos that evolves, taking us along for the ride. However, mankind must be in lockstep with the etheric force in order to evolve, or else Cosmos does it without us. The *Master of The Light* had shown me many esoteric – hidden historical – records of earth changes over hundreds of thousands of human years.

Velikovsky and many others have pointed out what really happened over the ages. They made it evident that this switching of consciousness had happened on this plane before, showing how in a rather short time frame, indigenous creatures of our planet would appear and disappear, as more highly-evolved life forms would take their place. The hidden key to our future is the possibility to leap beyond what was known before. A leap into the unknown. This transpired as we went through various sub-levels of density.

Now we are to break through to the other side of yet another level of consciousness, one where the mind would be free. The master of time explained to me that when a planet passes into a zone of higher vibration, a finer energy density prevails, so that the underlying spiral waves are much more complex, and thus the human and

465

other DNA structures simply evolve to much more complex or advanced levels, and much faster. It is a process all us are feeling now. However, those trapped in the lower density are responding to this change in negative ways.

While working with the trans-human souls I was shown that most of the visible *galactic dust* is not just dust, but that the galaxy from which we once emanated and of which essence we are now breathing, manifests the qualities of humanity. This new energetic DNA formation is taking effect throughout the whole galaxy. I realized that we are all moving onto a more evolved level of being via the *Zen of knowing*. As an astrologer I understand this in Cosmic and planetary star language, as well as how it applies to the earth.

This mystery which I know as *ancient beginnings*, or the path of the true masters, is our potential future as seen from the now. The function of the human being as a Cosmic architect is what my other world friends are all about. They say that we are their ancestors.

In our world they would easily have passed for space flight engineers; but in our journey to their sanctuary we spiraled to their inner world far below the lake, and into the core of our earth where their world is one of shining caves of crystals and zero-point instruments made of diamonds. There, they showed me a crystal millions of years old that drives the space ship we know as the earth.

I wondered what their aim had initially been, eons ago. Reading my mind, the time lord smiled and showed me the actual moment when they achieved the imagined dream of countless generations, the work of countless consecutive civilizations. It was what these men and women had been developing for what seemed an eternity – the unified field applications, with the light field and super defense technology of interstellar origin.

Anti-Gravity was redefined to propel planet Earth into a different direction. They came to this world to set mankind free. They made the invisible world visible for me. What I saw would fuse the mind of even the most advanced scientists. It was beyond present day human comprehension.

Entering Enlightenment

"He who exercises government by means of his virtue may be compared to the north polar star, which keeps its place and all the stars turn towards it."

~ Confucius

On the whole it had been a routine day of continued experimentation and other reality searching, one I was invited to witness and at other times take part in. The war was being waged by nations. Iraq was now divided in three parts; the names are the tags of our divided reality. Iran was caught up in the desert storm. Many thousands of years of betrayal and conflict flickered in the eternal flame of reality.

None of this was of interest to my hosts. They had completed the doorway to timelessness – the time loop – and now they were involved in manipulating events in another time zone that would lead to the transmutation of millions of human beings. I understood with compassion that this was to be for the good of this realm, should we decide to implement the final steps.

Due to the intricate and impossible to describe nature of reality, the time loop contained multiple dimensions or branches, of which they had to isolate one. They needed to select only one key element, as other key points of certain areas on earth served as a counterpoint to their connecting lines of time-space. Gridlines appeared, more advanced than we know them now, connecting the known and unknown, interconnecting lost and forgotten pyramids, mountain caves, other star gates, and time portals distributed on the globe, above and below sea levels. It took quite a while until all was ready. The time loop slid into place, and they saw a window opening as the central target faded from view.

A seer of their group told me there was a standard procedure for dealing with time loop distortions: They simply disconnect and begin again. The only problem is that there would be collateral damage should things go wrong. The engineers froze as they held the dimension doorway open. The first and secondary time loop dissolved, and the group of time lords I had joined stood in the room of crystal and light looking at what I would call *the sign*. We were looking at some octahedron-type star gate symbol rotating at high torque.

This portal of light, the octahedron, was a spiral vortex in motion. When still, it looked like the sign of a very ancient and timeless secret knowing within the esoteric, or the 11th harmonic level of the concord they were part of. Finding the sign seemed to have been their preoccupation. It was the goal they had set out to achieve countless generations before our time began. The engineers of the time lord locked onto the co-ordinates within the holographic crystal sphere. They established a connection, and the window opened for them, to see *beyond time*. The field opened before us. There it was: The Living Tao.

It was fascinating to behold the scene: I watched the whole reality unfold before them. The window of *the sign* opened up into what one could call *Nirvana* – the all that is and that is not. Human beings who embodied the essence of humanity stood before the most elusive door of reality – the entrance that shamans know as the path of

469

the soul: The way without a way, or the gateless gate. *Zen of Stars* is a pathless realm beyond dimensions, moving towards the whole. The time lords knew who they once were, who they are, why they exist and who they were to be in their *chosen future*.

I experienced with them their joy, the gratitude they felt for the harmonic of life that worked with them to move into that space. I experienced the peace and absolute purity of that moment. Unsought, unlooked for and totally unexpected, they had come to the point of their own beginning, seeing into the mirror of their trans-human soul, as a hyper-dimensional awakening no words can describe unraveled for all of us in that timeless space. A golden path stretched ahead of them – leading into eternity and beyond. The *passageway of the gods* had opened and all we had to do was walk through it.

Eons ago, a dis-resonant and fragmented force had corrupted the original stewards, or guardians of earth. It is a force we came here to undo. Our place is to be watchful that this thing does not ever again gain a foothold into our new society, or world. The highly strategic code of precision, by which the time travellers lived, had originally come from a pure source that in the beginning had known of the Tao-Te-Ching.

The knowledge of the Tao had come from a small band of men and women who had originally set out to free mankind. Of course, they had come from the stars. I do not recall what star that was, I imagine we are multi-level inter-planetary time travelers who call the whole universe their home. The world beyond shimmered into a passageway to another reality, one that silently waited for their approach. The lineage of the time lords had reached its destiny.

Imagine that you stand before the doorway of your own enlightenment and you cannot enter. The window of timelessness closed as I watched my friends embarking upon their long walk into the misty mountains, across the bridge of nowhere into a new time zone. The mists were rising on the lake and they would walk away, forever...

After a long and probing silence which seemed to penetrate the mountains around this lake, and after he had observed me seeing and being with the time lords, the *Master of The Light* asked me to convey to the people of the earth this one message:

"What you do you become, and what you become you do – to yourself and onto others."

He and I stood in the same power-place I liked to stand as a small boy, and we were gazing at the point where the Rhone River meets the lake of Geneva, and beyond that into the misty mountains, where elves and dragons come alive. He was well aware, and so was I that the same eyes looked at the same scenery, with some forty years of alchemy and wisdom in between. The soul enters and travels through this reality adventure, and as a protection he had shown me how not to invite fear and how to dispel evil.

To develop the path to *enlightenment*, the men and women of this world will have to turn around and undo all that has been done. It is a lot easier to pollute and damage a planetary eco-system, than it is to correct the damage once it has been done. The more messed up your life becomes, the harder it is to sort it out later down the road. Why dirty clean water when it is easier to keep it clean.

For centuries a few brave and wise souls asked their fellow travelers not to destroy the wilderness, the water resources, the seas the air and the delicate balance of the planet. They destroyed it regardless. World governments, with massive reserves taken from the individuals living within the system, have been amongst the largest investors in planetary destruction schemes, war being one example. But the mass of people enslaving their minds to the *ways of destruction* is the real death threat to the ecological and spiritual balance of the earth.

I could see it all laid out, like a badly drawn map superimposed on the ground of reality. And yet, I also saw ancient ancestors who are vibrantly alive in the land of spirit. They did not die, they are not dead and they wait for a point of entry into this world. When that point of entry comes we will learn to see things without the notion of time. Some of us already see this happening now, long before the invisible ones return.

With the third eye open, we will come to this threshold of perception, totally alert and without effort, to find out that our past is the future and that our future is the past. The so-called *occult*, which really means to make hidden, will return as science. Our birth right as Cosmic beings is in-divi-dual sovereignty, undivided from and part of creation.

We are here to co-create, and to cooperate intelligently with Cosmos – in this moment. For this to occur, we have to radically change our mind set, free our minds and connect to the trans-human or timeless soul within us.

The way we will enter enlightenment is through our own selfless gate, the timeless gateless gate within us.

The *Master of The Light* walked with me through the labyrinth of rainbow colors, where we would meet the other time lords within the vortex energy of the transformed reality field. They had undone the matrix of lower *vibrational hell*, and we met on the other side of the rip in the fabric of timelessness.

My awareness of, respect for and relationship with *The Master of Light* had undergone a complete transformation. The mirror of seeing had gained clarity. Saturnian authority of a timeless nature had merged with the kindness of our Geminian playfulness, we still had work to do and we intend to get it done.

Over the lifetime it takes a human star-being half breed to evolve, he had shown me through the maze of illusion created by thought – and from here on out it was the open sea of stars, of questioning and of a certain urgency to see things through. Intelligent compassion had become the way to lead the quest. The answers to the fundamental questions had been given. It was now only a matter of manifesting the future he had shown to me.

What is the meaning of enlightenment in the middle of planetary upheaval? He seemed to indicate to me that enlightenment had to be applied to be of use. I had never thought of enlightenment in terms of usefulness before, but now that I did and we were to compress the two realities into one, we would find a way just as the river moves from the mountain to the sea, by way of least resistance.

He and I saw a new movement on this planet. It was alive and flowering. The senile brain had come to the end of the road it had created for itself. Those *beyond survival* were ready for change. Our response to this movement of change would be complete and absolute.

* * *

Together we had reached the limits of thought and moved into the pure and limitless seeing. He and I knew we would now part company. No attachment needed to remain. No sadness permeated this moment, because at some point we will fuse again into one being. Then, from deep within, he communicated to my mind something that struck me in my core being, despite his incandescent smile when he said it:

"The morning of the magicians has dawned on you now. Always pay attention, because my change of worlds will come at any moment, and when that happens you must outlive me."

Man of The Future

"First they ignore you, then they laugh at you, then they fight you, then you win."

~ Mahatma Gandhi

In the Celtic world truth is stranger than fiction, and reality is even stranger. For the Celts Cosmos is mysterious and beyond our comprehension, and yet Cosmos communicates with us in a sign language we can understand. This sign language is what we call *reality* – this world – the universe, stars and planets.

The ancient Celtic way of perceiving and interacting with the unknown (life as a whole), is still alive within the DNA light strands that are waiting to be activated by a profound signal from the inhabiting souls – namely those of us on the earth today.

The Celtic mind is aware of circles of reality around and beyond the obvious physical sphere of continents, earth, planets and stars. St.Clair calls the expanding circles of reality the *invisible worlds*. For him they are also expanding circles of awareness. The greater your awareness, the greater your vision, and the greater your vision the greater your seeing.

In writing his profound and definitive *Zen of Stars*, St.Clair is "pointing the way", in a language that leaves the reader to expand their own awareness and their own understanding of events, in this world and the realms beyond. The invisible realms exist, and St.Clair is suggesting we open the door to their existence.

St.Clair, a prodigy child, was born in Zurich, Switzerland on 28 February 1959. He is the foremost strategic astrophysicist with a vast and highly accurate record of economic, social and political predictions. As a young man he studied law and political sciences at Zurich University and also served as a Swiss army officer – assisting a general – before beginning his career advising high-profile clients in special situations world wide.

In 1999, following twenty years of developing his skills of advisory expertise, and while continuing extensive esoteric and economic studies, he created a stunning twin work of cosmology and metaphysics. St.Clair's *Atlantis Oracle & Icons of Destiny*, while designing a unique understanding of mundane astrology and divination – delineate the future of humankind.

In 2002 St.Clair aired nine highly popular radio shows in Palm Beach, called *Passage11* on *WBZT*, and after being interviewed by Jeff Rense and Lou Gentile on their respective radio shows in 2004, St.Clair returned to his native Switzerland. His web site *Passage11.com* is consulted daily by thousands of visitors across the world. Over the years he has tirelessly worked to help people gain balance and peace of heart. He has often applied this high level of caring to extract clients and friends from the illusions of society.

The 21st Century has quietly swept to its surface a new visionary, and one of the most creative minds alive today, an astrophysicist who speaks of the most fundamental issues facing our humanity. The origin of his work is a mystery; brought to the public by a source St.Clair calls his ET Guidance. *Zen of Stars* is a spiritual revelation, narrated as a soul-expanding adventure revealing the profound discoveries that await humanity, as it considers our true Cosmic history. St.Clair's vision will unfold as real events, and validate its existence when hidden knowledge is uncovered by future explorations.

St.Clair's core message is: 2012 will see the merging of new realities in which you will be faced with direct action – not thoughts and words. He is saying: In essence, the world we live in today is a disconnection from our true past. We are psychic beings inhabiting a physical reality. The physical world is the vehicle for movement through time and space, but the psychic is the journey.

During his symposiums, St.Clair teaches awareness, transformation and intent. He urges you to access the silent knowing of your own spiritual sovereignty, and open the door to the unknown.

~ Susan McCulloch

Selected Reading – by St.Clair

Louis Pauwels & Jacques Bergier, *Le Matin Des Magiciens*, Paris, 1960

Andrews George C., *Extra-Terrestrial Friends And Foes*. Lilburn, GA: IllumiNet Press. 1993

Bach Richard, *Illusions – The Adventures of a Reluctant Messiah*. London: Pan Books Ltd. 1977

Brown, Jr. Tom, *Grandfather*. New York: Berkley Books. 1993

Childress David Hatcher, *Anti-Gravity & The World Grid*. Kempton, IL: Adventures Unlimited Press. 1987

Cremo Michael, *Human Devolution: A Vedic Alternative To Darwin's Theory*. Torchlight Publishing. 2003

De Chardin Pierre Teilhard, *The Phenomenon of Man*. Perennial. 1976

DeMeo James, *The Orgone Accumulator HandBook*. *Natural Energy Works*. 1989

Donovan Michael, *Letters Upon The Mast*. Hyannis, MA: Mapmaker`s Studio. 1995

Jones Marc Edmund, *The Sabian Symbols in Astrology*. Boulder, CO: Shambhala Publications. 1953

Jung Carl G., *Man and His Symbols*. London: Aldus Books Ltd. 1964

Kharitidi Olga, *Master Of Lucid Dreams*. Charlottesville, VA: Hampton Roads Publishing. 2001

Krishnamurti Jiddu, *Krishnamurti's Notebook*. Brockwood Park, Hampshire: Krishnamurti Foundation Trust Ltd. 2003

McLuhan T.C., *Touch The Earth – A Self-Portrait Of Indian Existence*. Oxford: Abacus.1971

Mitchell Edgar, *The Way Of The Explorer*. New York: Putnam Publishing Group. 1996

Moon R.G, *The Vortex Theory*. Fort Lauderdale, FL. 2003

Nhat Hanh Thich, *Calming The Fearful Mind – A Zen Response To Terrorism*. Berkeley, CA: Parallax Press. 2005

Polich Judith Bluestone, *Return of the Children of Light*. Rochester, VT: Bear & Company. 2001

Sitchin Zecharia, *The 12th Planet*. Rochester, Vermont: Bear & Company; Reprint Edition. 1991

Summer Rain Mary, *Phoenix Rising: No-Eyes' Vision Of The Changes To Come*. Norfolk, VA: HamptonRoads. 1993

Tesla Nikola, *The Fantastic Inventions Of Nikola Tesla*. Stelle, IL: Adventures Unlimited Press. 1993

Tolkien J R R, *The Lord of The Rings*. London: George Allen & Unwin Ltd. 1954

Trungpa Chögyam, *Cutting Through Spiritual Materialism*. Boulder, CO: Shambhala Publications, Inc. 1973

Tyson Donald, *New Millennium Magic*. St.Paul, MN: Llewellyn Publications. 1996

Velikovsky Immanuel, *Worlds In Collision*. Buccaneer Books. February 1996

Web sites of interest:
    WingMakers.com
    Thunderbolts.info

"St.Clair's magical work and existence is like the North Star. He is a guide and can lead you to your fullest potential – be it financial success, love, or peace – while steering you away from your darkest fears. The world is full of darkness and light... let this book show you your way to the light."

~ Kelly Leary, M.S.
(Granddaughter of Sir William Hill)

"Just wanted to thank you for making the sleeping seeds of light in me reawake through your writings..."

~ Eriksson Albin

"I have just recently returned from Russia where I was invited to speak at an international conference on nuclear physics. Anyway, when I came back I started thinking about you, and the predictions you made: that George Bush would win the election and that the ultimate theory of the universe would be discovered. It appears that you are exactly right."

~ Russel Moon, "Proving The Vortex Theory"

"St.Clair's predictive skills could save a government its intelligence service budget. His knowledge of astrology is astounding. He has been correct in his predictions on more occasions than I care to remember. His book's prophecies are haunting and should be on every world leader's desk. We would be well advised to listen to St.Clair."

~ Brad Robinson, President of The Millennium Group

"If Dane Rudhyar was *The* astrologer of the 20th century, then St.Clair is *The* astrologer of the 21st century."

~ Linn Ciesla

"St.Clair has combined his psychic recall, his command and knowledge of the history of our world, and his precise astrological calculations to bring you profound insights into the future of humanity. This work drives the force of futuristic thought and calls us to look deeper into our role in Earth's destiny. *St.Clair's gifts are Prophetic as he was born with Mars in Gemini by 'Arachne' by the Ancients.*"

~ Rev. Aphrodette North, PhD

"I would like to comment on your latest book... as well as your other writings: They have enabled me to keep my sanity. Thank you for confirming what many of us know in our hearts is true. Keep up your good work and we will do all we can to support your efforts from our end."

~ Rod Palmer

"I wanted to tell you what a deep impression this book made on me. For years, I've been reading all kind of religious, metaphysical or philosophical works, not in a very orderly way but it is all there, the basic knowledge. Now everything I wanted to know I have in just one book. What I liked the most is the underlying feeling of compassion & empathy for all things human & the desire to give active help for understanding the changing times ahead. Thank you for moments of undiluted intellectual & spiritual bliss. Sos lo mas!"

~ Marta Pont

Strategic astrologer & cosmologist St.Clair is known from many radio shows, and from his popular website: **www.passage11.com** – featuring his work; St.Clair accurately predicts global economic trends and world political events. His horoscopes and forecasts cover our time beyond 2020 – and into 2048. In this book of art and fiction he addresses what he terms our relationship with the invisible world.

You are your own future.

It is inside of you.

Access it!

www.zenofstars.org

Printed in the United States
75744LV00003B/68